REVIEW OF _____
THE U.S. GEOLOGICAL SURVEY'S
VOLCANO
____ HAZARDS PROGRAM

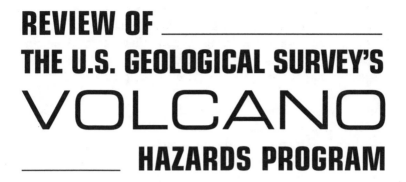

Committee on the Review of the USGS Volcano Hazards Program

Board on Earth Sciences and Resources

Commission on Geosciences, Environment, and Resources

National Research Council

NATIONAL ACADEMY PRESS
Washington, D.C.

NOTICE: The project that is the subject of this report was approved by the Governing Board of the National Research Council, whose members are drawn from the councils of the National Academy of Sciences, the National Academy of Engineering, and the Institute of Medicine. The members of the committee responsible for the report were chosen for their special competences and with regard for appropriate balance.

This study was supported by the U.S. Geological Survey, Department of the Interior, under assistance award No. 98HQAG2218. The views and conclusions contained in this document are those of the authors and should not be interpreted as necessarily representing the official policies, either expressed or implied, of the U.S. government.

International Standard Book Number 0-309-07096-1

Additional copies of this report are available from:

National Academy Press
2101 Constitution Avenue, N.W.
Box 285
Washington, DC 20055
800-624-6242
202-334-3313 (in the Washington metropolitan area)
http://www.nap.edu

Cover: Eruption of Mount St. Helens, courtesy of InterNetwork Media.

Printed in the United States of America

THE NATIONAL ACADEMIES

National Academy of Sciences
National Academy of Engineering
Institute of Medicine
National Research Council

The **National Academy of Sciences** is a private, nonprofit, self-perpetuating society of distinguished scholars engaged in scientific and engineering research, dedicated to the furtherance of science and technology and to their use for the general welfare. Upon the authority of the charter granted to it by the Congress in 1863, the Academy has a mandate that requires it to advise the federal government on scientific and technical matters. Dr. Bruce M. Alberts is president of the National Academy of Sciences.

The **National Academy of Engineering** was established in 1964, under the charter of the National Academy of Sciences, as a parallel organization of outstanding engineers. It is autonomous in its administration and in the selection of its members, sharing with the National Academy of Sciences the responsibility for advising the federal government. The National Academy of Engineering also sponsors engineering programs aimed at meeting national needs, encourages education and research, and recognizes the superior achievements of engineers. Dr. William A. Wulf is president of the National Academy of Engineering.

The **Institute of Medicine** was established in 1970 by the National Academy of Sciences to secure the services of eminent members of appropriate professions in the examination of policy matters pertaining to the health of the public. The Institute acts under the responsibility given to the National Academy of Sciences by its congressional charter to be an adviser to the federal government and, upon its own initiative, to identify issues of medical care, research, and education. Dr. Kenneth I. Shine is president of the Institute of Medicine.

The **National Research Council** was organized by the National Academy of Sciences in 1916 to associate the broad community of science and technology with the Academy's purposes of furthering knowledge and advising the federal government. Functioning in accordance with general policies determined by the Academy, the Council has become the principal operating agency of both the National Academy of Sciences and the National Academy of Engineering in providing services to the government, the public, and the scientific and engineering communities. The Council is administered jointly by both Academies and the Institute of Medicine. Dr. Bruce M. Alberts and Dr. William A. Wulf are chairman and vice chairman, respectively, of the National Research Council.

COMMITTEE TO REVIEW THE VOLCANO HAZARDS PROGRAM OF THE U.S. GEOLOGICAL SURVEY

JONATHAN H. FINK, *Chair*, Arizona State University, Tempe
CHARLES B. CONNOR, Southwest Research Institute, San Antonio, Texas
W. GARY ERNST, Stanford University, California
RICHARD S. FISKE, Smithsonian Institution, Washington, D.C.
CATHERINE J. HICKSON, Geological Survey of Canada, Vancouver, British Columbia
HARRY KIM, Hawaii County Civil Defense Agency, Hilo
STUART A. ROJSTACZER, Duke University, Durham, North Carolina
PAUL SEGALL, Stanford University, California
JOHN STIX, McGill University, Montreal, Quebec, Canada
FREDERICK J. SWANSON, U.S. Forest Service, Corvallis, Oregon

NRC Staff

TAMARA L. DICKINSON, Study Director
REBECCA E. SHAPACK, Research Assistant
JUDITH L. ESTEP, Administrative Assistant (through January, 2000)

This report has been reviewed by individuals chosen for their diverse perspectives and technical expertise in accordance with procedures approved by the NRC's Report Review Committee. The purpose of this independent review is to provide candid and critical comments that will assist the authors and the NRC in making their published report as sound as possible and to ensure that the report meets institutional standards for objectivity, evidence, and responsiveness to the study charge. The content of the review comments and draft manuscript remain confidential to protect the integrity of the deliberative process. We wish to thank the following individuals for their participation in the review of this report:

Grant Heiken
Los Alamos National Laboratory
Los Alamos, New Mexico

George M. Hornberger
Unviersity of Virginia,
Charlottesville

Donald Hull
Partners for Loss Prevention
Portland, Oregon

R. Wally Johnson
Australian Geological Survey
Canberra, Australia

Peter Mouginis-Mark
University of Hawaii at Manoa
Honolulu

John Pitlick
University of Colorado
Boulder

Michael Sheridan
State University of New York,
Buffalo

John Trapp
U.S. Nuclear Regulatory
    Commission
Rockville, Maryland

David Walker
Columbia University
Palisades, New York

Although the individuals listed above have provided many constructive comments and suggestions, responsibility for the final content of this report rests solely with the authoring committee and the NRC.

Contents

Executive Summary

Volcanic eruptions create some of nature's most dramatic displays. Depending on their magnitude and location, they also have the potential for becoming major social and economic disasters. To date, the United States has been relatively fortunate in this regard, having had only one large eruption near a major metropolitan area, that of Mount St. Helens near Portland, Oregon, in 1980. This event killed more than 50 people and caused considerable damage to infrastructure and timber resources. However, its most severe effects were restricted to lightly populated portions of rural Washington State. More frequent, recent eruptions in Alaska and Hawaii generally have had only local impacts.

Three recent developments make volcanoes increasingly dangerous for American citizens. First, rapid population and economic growth in the northwestern United States places more and more people, and some of our most critical industries closer to the regions' major sleeping volcanoes, including Mount Rainier and Mount Baker near Seattle-Tacoma, and Mount Hood near Portland. Second, the most heavily traveled transpacific air routes pass over more than 100 active Alaskan and Russian volcanoes, putting more than 10,000 people and millions of dollars worth of cargo in danger every day. A sudden eruption of ash, if undetected, could easily bring down a 747 by coating its engines with a debilitating layer of molten glass. Finally, the rapid and pervasive globalization of the economy means that U.S. companies and financial markets are increasingly vulnerable to disruptions caused by volcanic disasters anywhere in the world.

In the face of these dangers, federal and state government agencies in the United States rely primarily on the U.S. Geological Survey's (USGS) Volcano Hazards Program (VHP) to keep track of the status of all domestic volcanoes. This is a complex task that starts with fundamental *research* on the processes controlling the way volcanoes erupt. Research provides the basic concepts that underlie the various methods of volcano data collection and interpretation. The oversight continues with three operational components—*assessment* of hazard based on past history, *monitoring* of early warning signals that can indicate incipient eruptions, and design of *crisis response* strategies when large eruptions take place. Assessment's key challenge is deciding which volcanoes to study and in how much detail. Monitoring requires the measurement of geophysical, geodetic, and geochemical parameters, as well as baseline observations that allow premonitory changes to be recognized. A successful crisis response is characterized by rapid deployment of staff and equipment and by clear communication with civil defense officials and the public at large. A final *outreach* stage is necessary to inform civil defense officials and the public at large about the risks they face from volcanoes.

The VHP also maintains a team of scientists and technicians, partly supported by the U.S. Agency for International Development (USAID), that assists foreign governments with volcano hazard mitigation. This Volcano Disaster Assistance Program (VDAP) has had more than a dozen successful deployments in the past 10 years, most notably helping colleagues in the Philippines during the massive Mount Pinatubo eruption in 1991. This intervention saved tens of thousands of Filipino lives and hundreds of millions of dollars' worth of hardware at two U.S. military bases.

To assess how well the VHP is carrying out its mandated functions, the Geologic and Water Resources Divisions of the USGS requested in 1998 that the National Research Council (NRC) conduct an independent review. The NRC formed a committee of 10 members, representing industry, academia, and county and federal agencies, to evaluate how well the VHP fulfills these obligations. The specific charge was to answer two questions:

1. Do the activities, priorities, and expertise of the VHP meet appropriate scientific goals?
2. Are scientific investigations and research results throughout the VHP effectively integrated and applied to achieve mitigation?

Four meetings were held in 1999, during which the committee interviewed a variety of external experts, stakeholders, and USGS scientists, technicians, and administrators from both inside and outside the VHP. The committee also received written input from eight others. This report contains the findings of the committee.

In attempting to answer these two questions, the committee found that today's VHP is in many ways a product of its history. Throughout most of its existence, the program's focus was concentrated in Hawaii, which commonly has a limited range of relatively benign volcanic eruption styles. Hence the expertise required of VHP personnel, the types of monitoring and assessment activities, and the appropriate scientific goals were all somewhat restricted. Prioritization among these activities was straightforward and effective. However, the eruption of Mount St. Helens in 1980 presented the program with a greatly expanded set of scientific problems and mitigation activities, requiring a wider complement of expertise.

To acquire these new skills, the VHP depended upon retraining existing Geologic Division (GD) personnel and the addition of new hires who were mostly housed in the Water Resources Division (WRD). This response produced a more diverse but somewhat bifurcated staff. Most GD scientists had backgrounds in petrology, geochemistry, and sedimentology; relied largely on mapping and age dating as their primary tools; and preferred to work independently for extended periods. The generally younger WRD staff members included hydrologists, geophysicists, and structural geologists who were more familiar with quantitative methods, laboratory simulations, and mechanical modeling and who had greater experience with collaborative approaches. Furthermore, the two divisions had different policies and procedures for performance evaluations and job assignments. Melding these two components into a coherent organization capable of carrying out all of the VHP's responsibilities has been a major administrative challenge.

Among the many impressions that emerged from this review, two stood out—one positive and one negative. On the positive side, the VHP is comprised of a dedicated scientific and technical staff that has a wealth of practical experience, coupled with good theoretical understanding of underlying volcanic and hydrologic processes. On the negative side, an almost total failure to hire more than a token number of new personnel over the past 15 years has created a crisis of continuity in which much of the program's accumulated knowledge is in danger of being permanently lost due to upcoming retirements. Such a loss would have severe consequences during future volcanic emergencies. Because eruptions are so idiosyncratic and variable, the most valuable asset in assessing how a crisis will evolve is firsthand experience with previous events.

The VHP is at a major crossroads. Early in its history, scientific investigation formed the foundation of all of the program's activities. Its staff included some of the most accomplished volcanologists, petrologists, geochemists, and geophysicists in the country, if not the world. Attractive career paths in the VHP lured many of the most promising young geoscience graduates away from academic positions. All VHP scientists were expected and encouraged to carry out fundamental research as a central part of their jobs. Although much of this work was applicable to hazard mitigation, the connections were sometimes indirect. Responding to administrative redirection, basic research has become a lower priority for many members of today's VHP than the main mission of reducing the impact of volcanic eruptions. As a consequence, the center of gravity of volcanological knowledge, at least in the United States, has shifted away from the USGS and the VHP. Nonetheless, the VHP has made major contributions in a number of important areas: the development and application of assessment and monitoring techniques, crisis assistance, fluvial process knowledge, and aviation safety.

The implications for the VHP of the paucity of recent hiring, the prospects of flat budgets, and the shifting of staff emphasis from research to application are all the same—the program has to find better ways to leverage its resources so that it can continue to meet its mandate to mitigate volcanic hazards. The most direct way to accomplish this goal would be to hire more people. Alternatively, the VHP has to engage more actively in partnerships with other federal agencies, international counterparts, the private sector, and universities. Partnerships can take a variety of forms, from formal collaborations with other research groups on specific volcanological problems, to collocation of facilities with universities, to sabbatical programs through which VHP personnel spend time working with counterparts in other organizations, to the creation of volcanological grant programs jointly administered by the VHP and other agencies. Another practical way to respond to this deficit of new skills is for VHP management to aggressively promote the retraining of existing personnel.

A related issue concerns the way in which the VHP carries out hazard assessments. A common approach within the VHP has been for an individual scientist to have full responsibility for carrying out a long-term mapping and age-dating investigation of a single volcano. These projects can last for a decade or longer, during which time other scientists, both inside and outside the VHP, are discouraged from taking on similar studies at the same site. Preliminary data and conclusions may remain inaccessible until the entire work is completed. The result is that even rudimentary assessments for these volcanoes are unavailable, and the reports that eventually come out may have very personal stamps.

A newer and more efficient approach that has been successfully used by the VHP at Mount Rainier is to have a more coordinated effort through which a team of volcanologists evaluates many different aspects of the hazards in a shorter amount of time. This method brings a much broader set of expertise to bear on the specific assessment, including collaborators from outside the VHP. The faster turnaround time means that revised versions of hazard analyses are completed and made available to the public in a more timely fashion.

The autonomy shown by individual VHP scientists in establishing deadlines for their hazard assessments also has been reflected in the way volcanoes are selected for study. The committee was not presented with evidence of any long-term plan that indicated which volcanoes would be analyzed in what time frame. Rather, senior geologists in the program seem to pick the volcano they want to work on based on their own judgment and preferences. Although some oversight is provided today by USGS review panels that approve funding for individual research projects, the committee felt that the program manager, team leader and scientists in charge of each observatory could and should exercise more influence over the assignment process, consulting with each other to ensure a more coordinated approach to assessing the nation's volcanoes.

Among the volcanoes that have not been studied to date, the Aleutians were seen as the most problematic. The dangers posed to aircraft by sudden ash discharges are among the most serious threats to life and property overseen by the USGS. Although a program of lengthy and comprehensive assessment of all of the active Aleutians is not considered practical, preliminary studies for all of the potentially eruptible centers should be embarked upon immediately. The efforts of the Alaska Volcano Observatory (AVO) in this regard are commendable. AVO's budget has to be maintained or increased to a level that allows these initial studies to be carried out promptly, in order to guide the placement of instruments that can give early warnings to pilots of commercial and military aircraft.

More centralized organizational control is also essential for data collection, documentation, public access, and storage. Volcanology in general, and the VHP in particular, have not embraced the widespread movement toward universal and prompt data access found in other disciplines such as meteorology, seismology, and oceanography. In these fields, original data are commonly posted on the Internet in near real time, using widely accepted standards. In volcanology, there is a danger that the press or civil authorities might misinterpret prematurely released premonitory information, leading to inappropriate evacuations or panic. On the other hand, putting such data on-line allows both public education and more effective collaboration with scientists in other organizations. Overall, the committee feels that the advantages of timely access to data collected as part of the VHP's monitoring function outweigh the liabilities. Furthermore, the USGS should take the lead in establishing standards for archiving publicly accessible volcanological information and should evaluate which legacy data sets collected by observatories and individual scientists ought to be preserved and made accessible with defined standards for metadata and data quality.

Partnerships have the added advantage of allowing the VHP to fill in its deficiencies in expertise. For instance, gas geochemistry and remote sensing were identified as two important tools in modern volcanology that have been relatively neglected within the program. Universities and government facilities, such as National Aeronautics and Space Administration's (NASA) Jet Propulsion Laboratory and Goddard Space Flight Center contain many leaders in these disciplines. Better coordination with labs and universities would allow VHP staff members to take greater advantage of the latest advances in these fields.

The committee heard nearly universal praise for the outreach and educational activities carried out by the VHP, particularly by the Cascades Volcano Observatory. However, two bureaucratic barriers limit the effectiveness of these efforts. First, dissemination of the public education products generated by the VHP is very expensive to the program. Many recipients, such as schools and local governments, are willing to pay some or all of the costs of production. However, any revenue so collected goes into general federal government accounts rather than back to the program. This means that expanding outreach is discouraged by overall VHP budget limitations. A second barrier is that outreach activities have not received a high priority in VHP performance evaluations.

Today's Volcano Hazards Program evolved slowly from many antecedents within the USGS. For most of its first half-century, the VHP concerned itself with issues that had little impact on the American people. Population growth in the western United States, and the expansion of international commerce have greatly increased the threats posed by volcanoes, even though most citizens and policy makers remain unaware of these dangers. The size of the program has not kept pace with the growth of the risks. Although in its last major test, during the Mount Pinatubo crisis, the VHP was highly successful, the prognosis for the future is less optimistic. Diminishing staff sizes mean that the program is trying to do too many things with too few people. If existing trends continue, one of the next major eruptions will likely overwhelm its capability to respond.

The committee concludes that the VHP faces two choices: (1) its staff size must be significantly increased and/or better leveraged through partnerships; or (2) its mission must be scaled back in conjunction with major retraining of existing staff so that fundamental research, hazard assessment, and outreach play subsidiary roles to monitoring and crisis response. A combination of expanded partnerships with other research organizations and retraining of existing personnel could compensate for some of the lost capability associated with reduced staff size. However, the second and more drastic option of shrinking the scope of VHP responsibilities would likely be counterproductive in the long run. A VHP without a core research component will not be optimally prepared to negotiate the complex decision-making required in volcanic crises. In addition, it will not be able to attract and retain the best scientists, who will be sorely needed in a crisis situation.

Prologue 1

Mount Rainier Explodes!
Massive Eruption Devastates Seattle Area;
Officials Looking for Answers
Seattle, May 19, 2010 (AP)

Three months after rumbling back to life and exactly thirty years after the last major eruption in the lower 48 states, Mount Rainier exploded yesterday in a terrifying shower of ash, mud, and lava that took tens of thousands of lives and caused property damage initially estimated at more than 100 billion dollars. The unexpected magnitude of the eruption, which caught civil officials totally unaware, appears to have devastated the economy of a major Pacific gateway, setting the stage for a global financial crisis on a scale similar to that caused by the great Los Angeles earthquakes of 2003.

Yesterday's destruction was concentrated in three areas. The most immediate and deadly impact was felt in the southeastern suburbs of Seattle, where a searing blast of fine ash and gas flattened houses, factories, and bridges, and killed an estimated 13,000 people. Within an hour, mudflows surged down the volcano's western valleys, burying towns, highways, and railroads, and clogging the southern third of Puget Sound with logs and other debris. Throughout the day, the hills east of Kent were blanketed with up to 10 inches of rain-soaked ash, halting nearly all transportation and collapsing the roofs of thousands of homes and other buildings.

The devastation was not confined to the Puget Sound area. A 797 aircraft, en route from New York to Portland, crashed near Yakima when its engines shut down after passing through the ash cloud. All 650 people on board are presumed dead. The eruption's most concentrated economic effects appeared to be at aviation facilities in Renton, where wet ash caved in the roof of the main aircraft assembly plant, destroying 12 nearly completed supersonic transports valued at more than $3 billion each.

Also hard hit was the country's largest database server farm near Enumclaw, run by an industry-wide information technology consortium. A spokesperson said that backup systems should minimize the impact on global commerce, but investors' skepticism caused the stock price of several participating companies in the consortium to drop 40 percent overnight. Ironically, it was the inability of several Silcon Valley companies to respond quickly to the San Jose earthquake in 2007 that led to the relocation of these facilities to the Pacific Northwest in 2008.

Civil officials throughout the region were scrambling this morning to find scientists who could make sense of the disaster. Thirty years ago, the Volcano Hazards Program (VHP) of the U.S. Geological Survey provided official warnings that minimized loss of life from a similar eruption at Mount St. Helens. In 1991 the VHP and the Philippine Institute of Volcanology and Seismology forecast a massive eruption of Mount Pinatubo, saving tens of thousands of Filipino lives and hundreds of millions of dollars of military hardware at two nearby U.S. military bases. However, as part of the continued downsizing of the federal government in the early 2000s, the VHP changed from a scientific organization focused on research and prediction to a technical organization whose main tasks were printing maps and enforcing federal regulations. The Los Angeles disaster further shifted public attention and resources toward earthquake mitigation, leading to the complete shutdown of the VHP in 2007.

Contacted this morning at his home in Hilo, Hawaii, the retired former head of the USGS's Hawaiian and Cascades Volcano Observatories, sounded both angry and frustrated. "This is a human disaster that should have been avoided. Twenty years ago we had monitoring systems in place that could have tracked the movement of magma into the volcano, the weakening of the edifice's north side, the advance of mudflows toward Tacoma, and the paths of ash clouds. New techniques and sensors developed at universities in the United States, Japan, and Europe over the past decade could have further improved our predictive abilities, if we had people in place that knew how to use them. Instead we were lulled into a deadly complacency by leaders looking to save money and by Cascade volcanoes that typically awaken only once every century or two. Without a well-funded agency charged with coordinating volcano research and monitoring, this disaster is bound to be repeated."

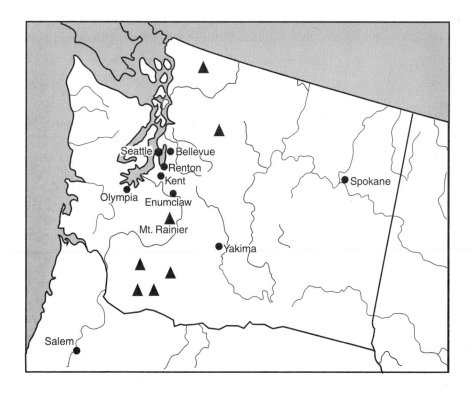

Figure P.1 Location map showing area affected by yesterday's eruption.

Prologue 2

Mount Rainier Erupts
Government Predictions Confirmed;
Major Disaster Averted
Seattle, May 19, 2010 (AP)

Mount Rainier exploded yesterday in a major volcanic eruption that caused serious property damage across the southern part of the Seattle-Tacoma metropolitan area. The relatively low initial death toll of 45 was credited to one of the most extensive disaster preparations in U.S. history, carried out over the past four years by state and municipal governments in conjunction with members of the United States Geological Survey's Volcano Hazards Program (VHP). Most of the casualties were thrill-seekers and photographers who ignored posted warnings and sneaked into the evacuation zone north and west of the volcano.

The eruption had three main parts, each affecting a different part of the region. At 8:31 a.m., the north flank of the volcano collapsed in a landslide, releasing a searing explosive blast that flattened trees, houses, and bridges across the southeastern suburbs of Seattle. Within an hour, mudflows surged down the volcano's western valleys where they were mostly contained and diverted by massive levees into huge, human-made basins. Throughout the day, the hills east of Kent were blanketed with up to 10 inches of rain-soaked ash that made roads impassable and collapsed the roofs of numerous buildings.

Although sobered by these property losses, public officials rejoiced that the unprecedented evacuations over the previous three days had been largely successful in preventing more deaths. The eruption was seen as a validation of the VHP, whose staff size increases, high-tech focus, and longstanding university and international partnerships allowed for near-pinpoint prediction of most of yesterday's events.

Early warnings from instrument networks and computer models let civil defense planners stage targeted evacuations, similar to those used since the 1980s to remove people from the paths of hurricanes. What was earlier feared to be a nearly impossible task—rapidly moving more than a million residents out of the potentially affected area in an orderly way—went remarkably smoothly, thanks in large part to a five-year public education campaign. Engineers were especially pleased that the widespread reinforcement and retrofit campaign for roofs, dams, bridges, and factories, carried out under the Federal Disaster Mitigation Act of 2004, appeared to have saved most of the structures.

Contacted at his home in Hilo, Hawaii, the retired former head of the USGS's Hawaiian and Cascades Volcano Observatories sounded both relieved and proud.

"Although the number of victims claimed by Mount Rainier was still too high, scientists and citizens alike should feel satisfied that their investments of time and money were rewarded in such spectacular fashion. The relatively small amount of destruction is a testament to the central role that federally coordinated scientific research can play in reducing the dangers of natural hazards."

For more details see Chapter 6.

1

Introduction

CONTEXT

The United States has more than 65 active or potentially active volcanoes, more than those of all other countries except Indonesia and Japan. During the twentieth century, volcanic eruptions in Alaska, California, Hawaii, and Washington devastated thousands of square kilometers of land, caused substantial economic and societal disruption and, in some instances, loss of life. More than 50 U.S. volcanoes have erupted one or more times in the past 200 years (Figure 1.1). Worldwide, approximately 550 volcanoes have had historically documented eruptions; at least 1,300 have erupted in the Holocene (past 10,000 years) (Simkin et al., 2000).

Eruptions do not affect only the area immediately surrounding a volcano. Ashfall can devastate areas downwind; ash clouds from major explosive eruptions can threaten large numbers of aircraft along well-traveled flight paths especially across the North Pacific; and large eruptions influence global climate and agricultural production. Recently, there have been major advances in our understanding of how volcanoes work. This is partly because of detailed studies of eruptions and partly because of advances in global communications, remote sensing, and interdisciplinary cooperation (Simkin et al., 2000).

The study of volcanoes is both empirical and probabilistic. Scientists look at what has happened in the past and attempt to predict what will happen in the future. The same could be said about this study of the U.S. Geological Survey's (USGS) Volcano Hazards Program (VHP): determining the future of the VHP requires a combination of careful observations and educated projections. The hypothetical news releases in Prologues 1 and 2 represent two strongly contrasting views of the ways the VHP might respond to a large volcanic eruption in the year 2010. If hiring patterns, budget decisions, and other trends that are rooted in current policies of the USGS and the Department of the Interior (DOI) continue, the results—as forecasted in Prologue 1—would be devastating. However, if the USGS is willing to make difficult changes in the VHP, then the number of casualties and the damage to the environment and the economy could be significantly reduced, as depicted in Prologue 2 and Chapter 6.

VOLCANO HAZARDS PROGRAM SETTING

The mission of the VHP is to "lessen the harmful impacts of volcanic activity by monitoring active and potentially active volcanoes, assessing their hazards, responding to volcanic crises, and conducting research on how volcanoes work" (USGS, 1997). The program has evolved in response to a variety of external pressures (see Sidebar 1.1). The first funding for studies of active volcanoes within the USGS budget was established in 1924 and went largely to support the Hawaiian Volcano Observatory (HVO) (Figure 1.1). From the 1920s through the 1970s, most USGS volcanic expertise was directed toward understanding Kilauea and Mauna Loa volcanoes, and HVO became the premier site for the development of volcano monitoring techniques and protocols (see Sidebar 1.2). During this same period, USGS scientists on the mainland examined prehistoric volcanoes and volcanic deposits throughout the western United States as part of other assignments: mapping mineral deposits, looking for geothermal areas, cataloging the natural resources of National Parks and wilderness areas, and characterizing fluvial processes.

The Robert T. Stafford Disaster Relief and Emergency Act of 1974 (Public Law 93-288) states that "the President shall insure that all appropriate federal agencies are prepared to issue warnings of disasters to State and Local officials." The director of the USGS, through the Secretary of the Interior, has been delegated the responsibility to issue disaster warnings "for an earthquake, volcanic eruption, landslide, or other geologic catastrophe." Therefore, the USGS has the responsibility to issue timely warnings of potential geologic hazards to the affected populace and civil authorities. The Stafford Act placed a new emphasis on the VHP's hazard mitigation role, which posed a challenge for the USGS.

SIDEBAR 1.1
Timeline of Events Affecting the USGS VHP

HVO established with non-federal funding	1912
Mount Katmai (Alaska) produces the most voluminous eruption of the twentieth century	1912
Private funding for HVO exhausted	1918
First federal support (U.S. Weather Bureau) of HVO	1919
U.S. Geological Survey support of HVO begins	1924
U.S. Geological Survey relinquishes control of HVO to National Park Service	1935
U.S. Geological Survey regains budgetary and management control of HVO	1948
Mauna Loa southwest rift eruption (Hawaii)—swiftly moving lava flows vividly demonstrate hazards	1950
Policy of rotating mainland scientists through HVO implemented	1956
First mention of Volcano Hazards Program in USGS budget	1968
Robert T. Stafford Disaster Relief and Emergency Act passed	1974
Mount St. Helens eruption (Washington)	1980
Establishment of Cascades Volcano Observatory	1980
Long Valley Caldera becomes restless (California)	1980
Establishment of Long Valley Observatory	1982
Growth of VHP and addition of Water Resources Division component	1981
Mauna Loa eruption (Hawaii)—lava flow threatens city of Hilo	1984
Nevado del Ruiz eruption (Colombia)	1985
Establishment of Volcano Disaster Assistance Program	1986
Internal review of USGS VHP chaired by Eugene Shoemaker	1986
Establishment of AVO	1988
Redoubt eruption (Alaska)—near fatal ash-aircraft engine interaction	1989-1990
National Research Council review of USGS VHP chaired by Meredith Ostrom	1990
Pinatubo eruption (Philippines)—VHP scientists contribute to successful hazards mitigation	1991
Merger with Geothermal Program	1995
Reduction in force in Geologic Division	1995
$3M budget cut in geothermal component	1996-1997
National Research Council review of USGS VHP chaired by Jonathan Fink	2000

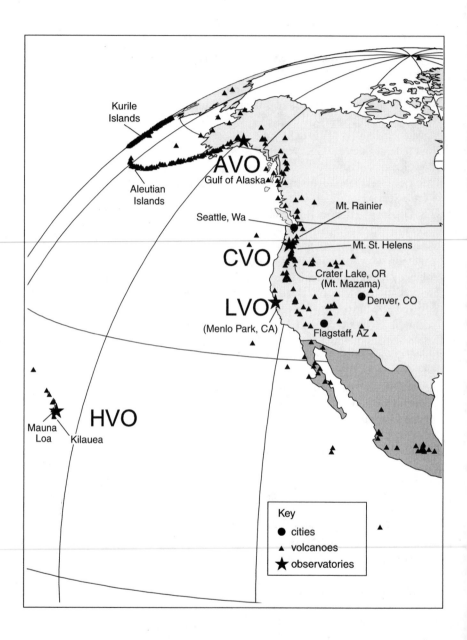

Figure 1.1 Map showing location of volcanoes and USGS volcano observatories.

SIDEBAR 1.2

An Early History of HVO and VHP

The first 30 years of HVO provided the scientific and philosophical foundation for the present USGS VHP. Founded in 1912 by Massachusetts Institute of Technology Professor Thomas A. Jaggar, Jr., and located at the summit of the active volcano Kilauea, the observatory quickly became a center of renown for volcano monitoring and hazards mitigation. The HVO logo shown above, whose motto can be translated, "No more shall the cities be destroyed," demonstrates the early commitment to volcano hazards mitigation, a major VHP objective today.

HVO flourished during its first five years, but Kilauea's gentle eruptions at the time were restricted to its remote summit crater, where no cities were threatened. In this time of relative quiescence, HVO staff studied molten basalt as it circulated in a summit lava lake, advancing knowledge of lava temperatures and gas content. Supported by private funds for its first six years, the observatory was operated by the U.S. Weather Bureau from 1919 to early 1924. The USGS took over management of HVO in June 1924, less than one month after explosions from Kilauea's summit showered nearby areas with falling debris, killing one person who ventured too close and reminding all that the volcano was potentially dangerous. Reduced Depression-era budgets forced the USGS to relinquish administrative control of HVO to the National Park Service in 1935, but an improved postwar economy enabled it to regain control in 1948, where it has remained to the present.

Hazards mitigation was HVO's number-one priority during the 28-year directorship of Jaggar, a volcano zealot who preached the virtues of volcano studies as the best way to protect life and property. Jaggar retired from the HVO directorship in 1940, and wartime HVO managers struggled to keep the organization afloat. Kilauea slumbered until 1952, but its giant neighbor volcano Mauna Loa erupted from its summit in 1949 and again from its southwest rift zone in 1950. The later eruption sent rivers of lava that flowed at velocities of up to 8 km per hour, cut a major highway in three places, and poured into the sea. The citizens of Hawaii received an abrupt reminder that their volcanoes pose serious threats to life and property. USGS management responded by increasing the scientific staff of the HVO and initiating volcano hazards research in the Cascade Range of the Pacific Northwest. In 1968, the "Volcano Hazards Program" budgetary line item was formalized.

Hawaii-based staff members had considerable experience dealing with volcanic activity, but this was for a limited range of eruptive styles. In contrast, the principal dangers to the western United States came from the explosive and effusive activity of stratovolcanoes and calderas, types of behavior not commonly observed in Hawaii. Thus, the sedimentologists, geochemists, and petrologists studying volcanoes in the western United States were asked to make predictions about future eruptions, even though the only, if any, activity most of them had seen was in Hawaii.

In many ways, the modern VHP was born when Mount St. Helens erupted violently on May 18, 1980. This was the first explosion of a mainland volcano since activity more than half a century earlier at Lassen Peak. The magnitude of the eruption and the range of its effects, especially the sector collapse and ensuing blast, were largely unanticipated, pointing out limitations in the expertise of VHP scientists, who had been drawn from both Hawaii and the mainland. Monitoring and describing this eruption gave a large number of geologists both inside and outside the USGS a tremendous boost in experience. Debris avalanches, pyroclastic airfalls, flows and surges, lava domes, and lahars could be studied in greater detail than ever before. The lessons learned at Mount St. Helens influenced the mapping of older volcanoes, emphasized the importance of process studies, and created a new appreciation among USGS scientists that exposure to as many active volcanoes as possible was an essential component of training for hazard mitigation at home.

The Mount St. Helens' eruption led to a large increase in funding for the VHP, as well as to the establishment of the Cascades Volcano Observatory (CVO) in nearby Vancouver, Washington (Figure 1.1). Because many of the most significant hazards of ice-clad stratovolcanoes involve fluvial processes, such as mudflows and debris flows, CVO included scientists from the USGS Water Resources Division (WRD), along with a roughly equal number of members of the Geologic Division (GD). Although many individuals from the WRD and GD cooperate on specific projects, especially during volcanic crises, members of the two divisions tend to work on different types of tasks. Many GD scientists focus on mapping, assessing, and monitoring the hazards of specific volcanoes, whereas WRD scientists are more apt to work on quantitative modeling particular geologic or hydrologic processes.

In 1985, Nevado del Ruiz volcano in Colombia had a relatively minor explosive eruption that melted a glacier and generated mudflows killing more than 23,000 people. Following this tragedy and building on the expertise gained at Mount St. Helens, the USGS and USAID set up the Volcano Disaster Assistance Program (VDAP), a well-equipped team of volcanologists and hydrologists able to respond to requests from foreign governments for technical and scientific assistance at the time of volcanic crises. Over the past 14 years, this relatively small team of scientists and technicians, based at CVO, acquired considerable experience through a series of deployments in the southwest Pacific, Africa, Latin America, and Alaska.

VDAP members' increased knowledge of the behavior of explosive volcanoes helped make possible the establishment of the Alaska Volcano Observatory (AVO) in 1988. AVO, a consortium set up by the USGS, the State of Alaska, and the University of Alaska, is charged with monitoring the active volcanoes of the Gulf of Alaska and the Aleutian Islands (Figure 1.1). Following a series of life-threatening encounters between jet aircraft and volcanic ash plumes in the late 1980s, the aviation industry recognized the risk posed by eruptions, especially along the heavily traveled North Pacific routes. AVO took on the role of providing timely warnings to airlines and pilots of any Alaskan eruptions capable of affecting aircraft safety. This responsibility has required AVO to gain better access to satellite-based remote sensing data, most of which originated from the National Oceanic and Atmospheric Administration (NOAA). It has also led to funding from the Federal Aviation Administration (FAA) for seismic monitoring of Aleutian volcanoes.

In addition to HVO, CVO, and AVO, the VHP has a team comprising a "virtual" Long Valley Observatory (LVO). Based at the USGS Western Regional headquarters in Menlo Park (Figure 1.1), LVO tracks volcanic unrest in and around Long Valley caldera in eastern California. Long Valley is the site of the most recent rhyolitic activity in the lower 48 states, when approximately 600 years ago, lava domes, pyroclastic flows, and surges erupted from a fissure more than 11 km long. The caldera itself formed in a major explosive eruption approximately 730,000 years ago. Seismic activity and extensive ground deformation beginning in 1980 caused concern among residents of the town of Mammoth Lakes, one of the largest ski and summer resorts in California. LVO staff members have worked closely with local officials in establishing strong lines of communication to maximize public understanding of the evolving volcanic situation and to minimize panic.

Budget History

The U.S. Geological Survey is the nation's largest integrated earth science agency and is the only science bureau in the Department of the Interior. The USGS is organized into four divisions (Geologic, Water Resources, Biological Resources, and National Mapping), with a total fiscal year 1999 appropriated funding level of $797.24 million. The VHP receives direct appropriations of $18.76 million (FY 1999). The overall funding for the VHP has increased slightly in real year dollars since 1978, but has remained flat since 1990 when adjusted for inflation (Figure 1.2), although responsibilities have increased in such areas as outreach, monitoring of Alaskan volcanoes, and international VDAP deployments. This decrease in buying power has caused significant hardship for the VHP and its ability to meet its public service mission today and especially in the future. The two peaks in funding in recent years trace to a supplement for the Redoubt eruption in 1990 and the merger of the Geothermal Program with VHP in 1995. The geothermal component of the combined program was cut substantially in FY 1996 and 1997, and much of the remaining geothermal project work was redirected to hazards studies.

The VHP's direct appropriations are supplemented by funds derived from other agencies under reimbursable contracts (Figure 1.3). In FY 1999, the VHP received $2.83 million of such funding, up substantially since 1995 (Figure 1.4). The VHP is funded jointly by the GD and WRD. In FY 1999, 68 percent of the VHP funding was spent within the GD, 24 percent within the WRD and 8 percent on cooperative agreements with state and university partners. These percentages have remained nearly constant since 1995 (Figure 1.5).

Staffing History

The VHP staff members are widely distributed in the western United States (Figure 1.1), working at three main observatories (Hawaiian, Alaska, and Cascades); one virtual observatory (Long Valley); and regional USGS offices in Menlo Park (California), Denver (Colorado), Flagstaff (Arizona), and Seattle (Washington). In 1999 the program had 135 full-time equivalents (FTE), with 35 in WRD and 100 in GD. Of these, 10 FTEs are supported by outside agencies under reimbursable contracts,

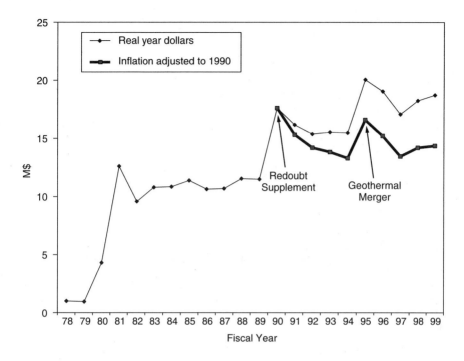

Figure 1.2 Appropriated funding level for VHP.

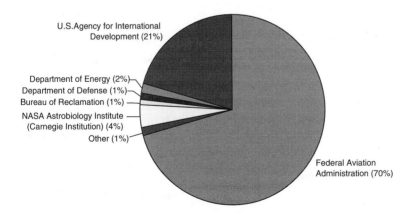

Figure 1.3 Reimbursable funding profile for VHP: income from outside agencies.

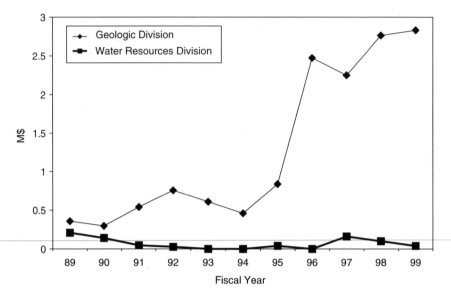

Figure 1.4 Reimbursable funding levels for GD and WRD components of VHP.

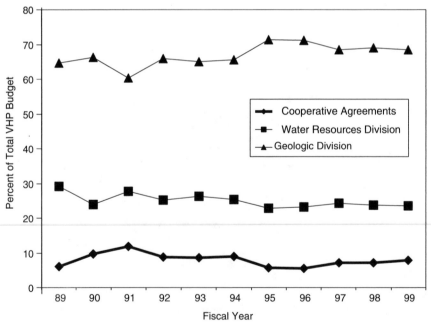

Figure 1.5 Budget activities within the VHP as a percentage of total funding level.

including 5 from the FAA and 4 from the USAID. The number of FTEs in GD has decreased significantly since 1995, whereas the number in WRD has decreased less dramatically (Figure 1.6). The GD has no data on program-staffing levels for the VHP before 1995. Prior to this time, staffing was accounted for by disciplinary branches, not by program funding sources.

Program Approaches

To help society prepare for and deal with the effects of volcanic eruptions, the VHP uses five interrelated approaches. (1) Long-term *hazard assessment* of the potential magnitude and timing of future eruptions requires documentation of a volcano's past activity. A key challenge in assessment is deciding which volcanoes to study and how much detail such studies should seek. (2) Volcanoes that experience ongoing or recent activity require the *monitoring* of signals that might portend an incipient eruption. Monitoring demands the establishment of geophysical, geodetic, and geochemical baselines that allow premonitory changes to be recognized. (3) When an erupting volcano provides a direct threat to society, the VHP establishes a *crisis response.* A successful crisis response is characterized by rapid deployment of staff and equipment and by clear communication with civil defense officials and the public at large. (4) In addition to performing assessment, monitoring, and crisis response, VHP staff members conduct topical studies of geologic processes that allow for better understanding of the causes and consequences of volcanic hazards. (5) A final essential function of the VHP is to communicate to the civil authorities and the surrounding communities the results of its studies.

These five approaches all aim to help society respond to the dangers posed by volcanoes. Another way to view these activities is to consider a continuum of three overlapping types of societal response to eruptions: (1) Research (knowledge acquisition), (2) Operations (knowledge application), and (3) Outreach (knowledge translation). Research provides the basic information and concepts that underlie the various methods of volcano data collection and interpretation. Until the 1980s, the VHP accounted for much of the fundamental research on volcanic activity in the United States. More recently, scientists from universities and

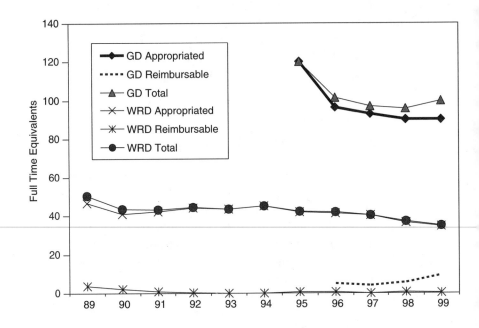

Figure 1.6 Staffing levels in GD and WRD components of VHP.

the private sector have increasingly shared these functions. *Operations* refers to all of the work involved with direct mitigation of eruptions— assessment, monitoring, and crisis response. The VHP continues to be the principal source of nearly all of these essential activities in the United States and, through VDAP, is a significant contributor to similar efforts around the world. *Outreach* includes interactions with communities surrounding volcanoes. These types of responsibilities have grown in recent years, both for the VHP and for other agencies, although the VHP's ability to maintain them has at times been hampered by insufficient funds. In evaluating the Volcano Hazards Program, perhaps the most basic question that USGS management must consider is, "What is the appropriate balance among research, operations, and outreach within the VHP, given changes in budgetary priorities and ancillary capabilities of other organizations?"

STUDY AND REPORT

To provide a fresh perspective and guidance to the VHP about the future of the program, the Geologic and Water Resources Divisions of the USGS requested that the National Research Council conduct an independent and comprehensive review. In December 1998, the Board on Earth Sciences and Resources within the Commission on Geosciences, Environment and Resources formed an ad hoc Committee to Review the U.S. Geological Survey's Volcano Hazards Program. The committee consists of 10 geoscientists and volcanology experts from industry, academia, and county and federal agencies. Its members have recognized expertise in physical volcanology, geochemistry of volcanic gases and rocks, igneous petrology, geophysics, groundwater hydrology, remote sensing, geodesy, eruption prediction, volcano hazard assessment, emergency management, preparedness and response, risk communication, and science administration. Brief biographies of the committee members are provided in Appendix A.

Using the Volcano Hazards Program 1998-2002 Science Plan (USGS, 1997) as a starting point, the committee was asked to examine how the program is adapting to changes that have occurred as it has grown following the eruption of Mount St. Helens in 1980. The charge to the NRC committee included the following questions:

- Do the activities, priorities, and expertise of the program meet appropriate scientific goals?'
- Are the scientific investigations and research results throughout the program effectively integrated and applied to achieve hazard mitigation?

The broad purpose of this examination is to provide fresh perspective and guidance to the VHP about its future directions.

A previous review of the VHP, chaired by Eugene Shoemaker, was conducted in 1986. That committee recommended, among other things, that (1) the level of effort for hazard assessment be increased; (2) seismic monitoring be conducted at several specific volcanoes; (3) an adequate professional staff in gas monitoring be maintained; (4) improved techniques for monitoring outgassing of CO_2 be explored; (5) new instrumentation (sensors on Department of Defense satellites and microbarographs) for monitoring Alaskan volcanoes be evaluated; and

(6) steps should be taken to ensure interdivisional focus of effort in VHP. In 1990, the NRC Committee Advisory to the U.S. Geological Survey conducted a review of the status of the programs within the VHP (NRC, 1990). That committee also commented on the lack of effective coordination between the efforts of the two divisions, recommending a single program coordinator, and encouraged studies of volcanoes worldwide for knowledge that can be gained and subsequently applied where needed.

The committee held four meetings between March and August 1999. These meetings included presentations from and discussion with staff of the VHP and other USGS programs, representatives from local, state, and federal agencies, and academia. The committee also received written responses to questions about the VHP from stakeholders. Individuals who provided the committee with oral or written input are identified in Appendix B. As background material, the committee reviewed relevant USGS and VHP documents and materials through August 1999, pertinent NRC reports, and other technical reports and published literature.

This report is organized around the three components of hazards mitigation. Chapter 2 deals with research and hazard assessment. Chapter 3 covers monitoring and Chapter 4 discusses crisis response and other forms of outreach conducted by the VHP. Chapter 5 describes various cross-cutting programmatic issues such as staffing levels, data formats, and partnerships. Chapter 6 offers a vision for the future of the Volcano Hazards Program, and Chapter 7 summarizes the conclusions and recommendations of the preceding chapters. Throughout the report, major conclusions are printed in italics and recommendations in bold type.

The committee has written this report for several different audiences. The main audience is upper management within the USGS and the VHP. However, the committee believes that scientists within the VHP will also find the report valuable. The report is written in such a manner as to be useful to congressional staff as well.

2

Research and Hazard Assessment

The ability of any organization or individual to translate observations of volcanoes into interpretations of present or future eruptive behavior requires grounding in the theoretical understanding of how volcanoes work. At the same time, the advance of theory depends on a steady infusion of new measurements. Thus, the interactions between the development of new concepts and volcanic data collection and analysis are inextricably intertwined.

Most of this report focuses on the daily activities of the VHP, what may be referred to as the "operational" component of its mission. In this chapter, the committee begins by briefly examining the changing contributions of VHP scientists to the development and promulgation of the theoretical framework of volcanology. *Hazard assessment,* the operational activity that most clearly connects with research, is then explored in some depth.

RESEARCH

It is difficult to separate the contributions to basic volcanological knowledge made by VHP scientists from those made by their colleagues in other parts of the USGS, other government agencies, universities, other countries, and the private sector. Nonetheless, throughout much of the second half of the twentieth century, members of the present day USGS Volcano Hazards Program were national if not global leaders in the formulation of ideas about how volcanoes work. Building upon a steady stream of fresh observations from the HVO, VHP personnel advanced the understanding of the roles and significance of earthquakes, deformation, explosivity, gases, and lava flow mechanics in the evolution of ocean island volcanoes. Other USGS scientists conducted long-term studies of individual volcanoes in the western United States, especially the Cascades Range. Many of these studies were carried out for purposes other than hazard assessment, such as the identification of geothermal and mineral resources and the geologic mapping of wilderness areas and national parks, but many findings were applicable to VHP goals.

The VHP has defined its scientific priorities for the next five years by the fundamental questions that must be answered to fulfill the program mission of effective mitigation and useful warnings (USGS, 1997):

- Where are potentially high-hazard volcanic areas?
- Where is volcanic unrest occurring and in what manner?
- Is a restless volcano going to erupt? When? How long and how dangerous will the eruption be? How will eruptive style change over time?
- How can the potential for short- and long-term volcano hazards potential best be communicated?

In FY 1999, the VHP funded more than 60 science projects, most of which directly or indirectly addressed these questions.

The committee did not review the individual VHP research projects nor did it conduct an in-depth assessment of the research component of the program. It started with an awareness of the outstanding reputation of much of the research carried out by the VHP. However, the committee heard several anecdotes about longstanding research projects that have questionable connections to the primary mission of the program. The committee feels strongly that USGS management must ensure that most, if not all, basic research projects are directed toward the above three priorities. Such assurance can come from stronger internal USGS programmatic oversight and from careful structuring and enforcement of the annual performance plans of individual research scientists. This oversight should include subjecting proposals for research projects to external peer review. Additionally, as emphasized elsewhere in this report, prompt publication of research findings is essential. The committee is aware of important research findings that have languished for years (or even decades) without being published. These problem situations must be addressed and solved.

One of the most important long-range issues that the VHP must face is deciding how central in-house basic research will be to its mission in the future. Such research is also being done at universities, government laboratories, and non-U.S. institutions. Thus, one might argue that the VHP could forgo its basic research activities without having a major impact on the state of knowledge of volcanic processes. On the other hand, eliminating this program element altogether would likely damage the intellectual vitality of the VHP and make it more difficult (if not impossible) for the program to hire top-flight young scientists. Furthermore, because most USGS scientists lack some of the other commitments of their academic counterparts, such as teaching and grant writing, they may be better able to pursue long-term research projects than university faculty and students. *The committee believes that if the VHP is faced with continuing budget shortfalls, it could elect to reduce fundamental research activities and redirect scarce resources to monitoring and crisis response functions, which it is uniquely positioned to do* (see Chapters 3 and 4). *However, these savings would come at a high cost.* The ability of the VHP to respond to volcanic crises would be compromised by a lack of expertise in hazard assessment or volcano process studies.

One possible solution would be for VHP members to collaborate more on research projects with scientists outside the USGS, particularly those from universities and laboratories of other government agencies. More active collaborations, coupled with an extramural grant program for academic researchers overseen partly or completely by the VHP, would help ensure that more investigations that are directly relevant to the program's mission would be carried out (see Chapter 5). In addition, a more proactive and vigorous approach to retraining of existing personnel could help maintain the breadth of expertise needed to understand and respond to volcanic behavior.

HAZARD ASSESSMENT

Most VHP scientists and technicians spend the majority of their time in operational activities that lie within continuum of assessment, monitoring, and crisis response. This section considers hazard assessment, which forms a foundation for the other two operational components. The committee first describes what volcano hazard assessment is and why it falls within the purview of the USGS. Next the different types of volcanic hazards are categorized. A status report on hazard assessment within the VHP is then presented, focusing first on three important approaches (mapping and dating, theoretical modeling, probabilistic methods) and then on how the different observatories carry out these responsibilities. Finally the committee considers the future of assessment within the VHP and recommends that greater emphasis be placed on prioritization, collaboration, and consistent data archiving in order to help the program carry out its mandate more effectively and economically.

What Is Volcano Hazard Assessment?

Volcano hazard assessment aims to determine where and when future volcano hazards will occur and their potential severity. This kind of appraisal provides a long-term view of the locations and probabilities of large-scale eruptions and related phenomena such as volcanic debris avalanches and tsunamis.

The boundaries between hazard assessment and basic volcanological research are indistinct. Maps of volcanic deposits can be used either to reconstruct the history of a particular volcano or to decipher eruption processes that occur at many volcanoes. Stratigraphic sequences of ash deposits can similarly reveal how one volcano has behaved, or they can provide a basis for comparison across an entire volcanic arc, leading to the discovery of fundamental principles.

Volcano hazard assessment involves a combination of three methodologies. First, volcanic deposits are recognized and mapped, and the associated materials are dated to provide a chronology of the related eruptive events. Second, laboratory and numerical simulations of the physical and chemical processes that initiate volcanic hazards are used to better understand and constrain their underlying causes. Third, information from both field-based studies and simulations are combined to make statistical assessments of the probability of future events. This work relies on one of the fundamental axioms of geology: the past is the key to the present and future. In other words, volcanoes with the most eventful history of activity are those most likely to erupt again in the future, and the style and nature of past eruptions are the best guides to future behavior.

Why Does the USGS Do Volcano Hazard Assessment?

Assessment is integral to the mission of the VHP. Under the Stafford Act (Public Law 93-288), the USGS has the responsibility to issue timely warnings of potential geologic hazards to the affected populace and civil authorities. A narrow interpretation of this law might restrict the USGS to monitoring only those volcanoes showing outward signs of an imminent eruption, ignoring assessment altogether. This temptation could come from the twin desires to save money and to emphasize only the most obvious needs associated with volcanic hazards. However, assessment is an essential complement to monitoring: it provides an important means of prioritizing which volcanoes should be monitored and which types of data should be collected. It helps determine the potential magnitude of imminent volcanic hazards and allows the public to be educated about the likely consequences of volcanic activity.

What Hazards Are Assessed?

Volcanic hazards are highly varied in nature, frequency, size, area of impact, and complexity (Figure 2.1). There are two basic types of eruptions: (1) effusive, which generate lava flows and domes, and (2) explosive, which produce mixtures of ash, blocks, and gas, known as pyroclastic flows, capable of traveling large distances at great speeds. In the presence of water, from groundwater, precipitation, lakes, streams, or melting glaciers, ash and other loose deposits may become mobilized into highly destructive debris flows, known as lahars. Ash, consisting of small volcanic particles, can be thrown kilometers above the volcano during explosive eruptions and may be carried as far as hundreds of kilometers downwind. When ash settles out of the atmosphere it leaves thick deposits whose weight may cause the roofs of buildings to collapse. Ash deposits can also seriously alter drainage patterns and sediment loads, leading to widespread flooding. Volcanoes themselves tend to be unstable structures, occasionally collapsing to form landslides and, rarely, massive volcanic debris avalanches. Debris avalanches can bury extensive areas or, when they enter bodies of water, lead to large water waves or tsunamis. Gases escaping from subsurface magma can lead to respiratory distress or death in humans and other animals and may destroy forests and crops.

The potential impact of eruptions is not restricted to destruction on land (Figure 2.2). One of the greatest and least appreciated volcanic hazards is the threat that ash clouds pose to aircraft. Nearly all nonstop airline routes from North America to Asia pass over (or short distances downwind from) tens of potentially active volcanoes in Alaska and Russia. Ash may cause engine failure even when it is too fine grained to be visible or detectable by airborne radar.

Not all volcano hazards are related directly to eruptions. Damaging mass movements can take place without any magmatic discharge, which occurred in 1888 when a large part of Mount Bandai, Japan, collapsed to form a debris avalanche. Giant submarine landslides have formed by partial collapse of nearly all of the Hawaiian volcanoes. One of these, originating from the flanks of Mauna Loa, generated a tsunami that washed more than 200 meters up of the slope of the neighboring island of Lanai. Damaging earthquakes may accompany the underground movement of magma, even if molten material does not erupt at the earth's surface. The extensive range of hazards that must be evaluated requires the combined knowledge of a broad array of scientists, including geologists, geophysicists, hydrologists, geotechnical engineers, atmospheric physicists, and statisticians. *Because assessment is inherently interdisciplinary, the VHP needs access to a diverse set of expertise, either within its own ranks or through collaborations with outside groups.*

What Is the Status of Assessment Within the VHP?

Traditionally, volcano hazard assessment within the VHP focused on field-based and geochronological studies of individual volcanoes. Initially geochemists and petrologists from the GD undertook many of these investigations for purposes other than analysis of hazards. Since the eruption of Mount St. Helens in 1980 and the creation of the CVO, hydrologists, geomorphologists, and sedimentologists from the WRD have also taken part in these appraisals, incorporating laboratory and numerical simulations and field studies of the processes that lead to debris flows and steam explosions. Creation of the AVO in the early 1990s allowed glaciologists, atmospheric scientists, and remote sensing specialists to play a role as well. Because the VDAP team can be called on to assist with volcanic crises anywhere in the world, it generally has

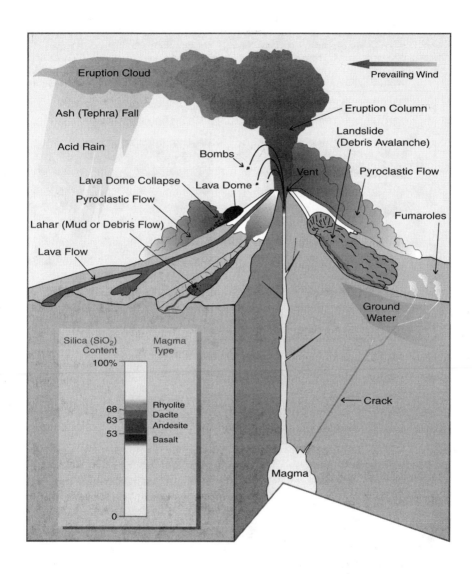

Figure 2.1 Simplified sketch of a volcano typical of those found in the western United States showing a variety of hazards associated with volcanoes (USGS, 1998). Graphic designed by Sara Boore, Bobbie Myers, and Susan Mayfield.

A. Structural Damage Due to Volcanic Ash
Heavy ash fall from Mount Pinatubo caused this
World Airways DC-10 airplane at Cubi Point Naval
Air Station to set on its tail. The ash cloud caused
eleven commercial aircraft emergencies. Photograph
by R.L. Rieger, U.S. Navy on June 17, 1991

B. Damage due to lahars
Within four hours of the beginning of the
eruption of Nevada del Ruiz, lahars had
traveled 100 km and left behind a wake of
destruction killing more than 23,000 people.
Hardest hit was the town of Armero, which
was located in the center of this photograph.
Photograph by J. Marso in late November
1985. Rio Lagunillas, former location of
Armero.

C. Effects on Structures
The Wahaula Visitor Center, Hawaii Volcanoes
National Park, was engulfed by a lava flow and burst
into flames (June 22, 1989). Note flow at left in photo.
Photograph by J.D. Griggs, Hawaiian Volcano
Observatory, U.S. Geological Survey.

D. Effects of Flooding Rivers
At Mount St. Helens, millions of tons of
debris broke loose and traveled down the
Toutle river increasing danger to the
communities downstream.
May 18, 1980

E. Damage due to Lateral Blast
The slope of Smith Creek valley, east of Mount St.
Helens, shows trees blown down as a result of the
blast. Photo by Lyn Topinka September 24, 1980

Figure 2.2 Impacts of volcanic eruptions.

had to rely on assessments carried out by organizations other than the VHP. Outside the USGS, assessment has tended to include a growing reliance on physical and probabilistic models. The following material describes some of the methods used in hazard assessment and their status within the VHP.

Mapping and Dating

Geologic mapping, stratigraphy, geochronology, and physical volcanology provide the backbone of volcanic hazard assessments by revealing trends in eruption timing, volume, and explosivity. Historically, the USGS has done an excellent job of incorporating these types of geologic data into its assessments. For example, insight gained from the geologic mapping of Mount Pinatubo, coupled with knowledge of magmatic and eruption processes, increased the accuracy of forecasts by the USGS and the Philippine Institute of Volcanology and Seismology (PHIVOLCS) and arguably reduced loss of life when this volcano erupted violently. Many USGS scientists first observed massive debris avalanches during the 1980 eruption of Mount St. Helens. Although not widely appreciated prior to this event, the potential for such hazards is now acknowledged in assessments for Mount Baker, Mount Rainier, and similar volcanoes. *The committee commends VHP efforts to integrate findings of geologic studies into volcanic hazard assessments.* An ongoing challenge is to quantify geologic data more effectively in ways that optimize their use in such assessments. The VHP has also benefited from having outstanding in-house instrumentation and the professional expertise needed for various kinds of age dating of geological materials. USGS management will have to evaluate whether it can continue to afford the associated high expense or whether it would be more appropriate to contract out such services to the private sector.

Modeling

Although mapping and dating of volcanic deposits can provide a good framework for hazard assessment, mechanical models of physical, chemical, and hydrologic processes help refine forecasts of the types and magnitudes of future eruptions. Both numerical models and laboratory simulations can relate the boundary conditions on a volcano to the likely consequences of any incipient eruptive activity. To date, this modeling work has focused on landslides, debris flows, lava flows, and various types of explosive eruptions. For instance, results of large-scale flume experiments carried out by CVO scientists at a unique facility in Oregon (e.g., Iverson, 1997) now allow them to calculate how far downstream and how rapidly a lahar will travel if it is triggered by the melting of a glacier perched above a specific volcanic drainage. Another approach, the so-called energy-line model (e.g., Malin and Sheridan, 1982), can determine the area likely to be covered by the fallout from an explosive eruption column based on the observed height of the plume. Recent laboratory and field-based studies of lava domes relate the textural and structural patterns observed on their active surfaces to the likelihood that they will explode (e.g., Fink and Griffiths, 1998).

Although there has been some VHP participation in the development of these models, especially those related to hydrologic and sedimentologic phenomena, most have been created by non-USGS scientists. VHP hazard assessments generally do not incorporate the latest of these methods. Besides potentially limiting the scope of such assessments, the lack of a strong theoretical focus within the VHP makes it more difficult to test these models thoroughly against real-time eruption data collected by the program's scientists. *The committee encourages the VHP to include more theoretical modeling of volcanic phenomena in its hazard assessments.*

Probabilistic Hazard Assessments

Because it is impossible to predict eruptive behavior with certainty, particularly for dormant volcanoes, most hazard assessments are inherently probabilistic in nature. For example, analysis of stratigraphic and radiogenic data may suggest that a given volcano has a 30 percent probability of erupting explosively in the next 50 years. Such a statement does not provide all of the detailed information needed by preparedness officials, which is where *conditional probabilities* become useful. For instance, if an explosive eruption occurs, what is the probability that ash accumulation will exceed some threshold? This approach enables volcanologists to consider a complicated series of events discretely or to estimate hazards based on empirical or subjective information that in practice may be incomplete or poorly understood. As another example, volcanic unrest often appears to build exponentially, and estimates of the probability of future eruptions may be made based on this kind of trend. However, these estimates must be conditioned by the likelihood that the exponential model is correct.

Use of these three approaches to hazard assessment—mapping and dating, theoretical modeling and probability calculations—by the VHP reflects the training of its participants. Most of the GD scientists have backgrounds in petrology, geochemistry, and field mapping and thus are most familiar with the more traditional stratigraphic and geochronologic methods. Many of the WRD scientists are experienced with simulations and mechanical modeling, which explains why some of the finest theoretical explanations of debris flows and sediment transport have been carried out by members of the VHP. Probabilistic approaches are relatively recent additions to the VHP assessment repertoire, but they are receiving more attention lately because of their obvious utility in communicating with civil defense authorities and the general public. *The committee strongly encourages the VHP to develop a balanced assessment program that takes advantage of the full range of techniques available to volcanologists today.*

The State of Volcano Hazard Assessment at USGS Observatories

Assessment priorities vary from observatory to observatory, reflecting local differences in the nature of the volcanic hazard and the expertise of resident scientists and technicians. Here the committee offers a description of the state of volcanic hazard assessment in each of the regions overseen by the VHP's observatories.

Alaska Volcano Observatory

In Hawaii and the Cascades, threats to nearby population centers are the focus of most volcanic assessment. By contrast, Alaska has small population clusters around its volcanoes. Thus, the most serious hazards associated with Alaskan eruptions are those that occur at a distance from the eruptive site. Tsunamis generated by volcanic landslides and earthquakes can potentially affect coastal areas throughout the Gulf of

Alaska, as well as elsewhere around the Pacific Rim. Additional threats result when eruptions melt glaciers that then generate debris flows capable of destroying various remote installations, such as logging and fish-processing facilities. Alaska has a rich and diverse fauna and flora. Species preservation plans, which now are partly the responsibility of the USGS (although not the VHP), may benefit from volcano hazard assessments.

Volcanic ash interaction with jet aircraft poses the greatest danger from Alaskan volcanoes, because ingestion of ash can result in engine damage or failure. On average, approximately four volcanoes per year erupt ash clouds of sufficient height and volume to endanger aircraft in the heavily traveled North Pacific air corridor. *Although responsibilities for monitoring and crisis response in Alaska are shared among the VHP, the National Weather Service (NWS), and the FAA, only AVO is capable of (1) establishing the historical context of future explosive eruptive activity, (2) providing advance warning of an impending eruption, and (3) conducting ground monitoring that can confirm an eruption is actually in progress.*

Because of the nature of these dangers, AVO has placed greater emphasis on monitoring and crisis response than on long-term hazard assessment. Only a few Alaskan volcanoes have even rudimentary hazard maps (Appendix C). The expense and logistical difficulties associated with access in Alaska preclude the kind of comprehensive mapping strategy carried out by CVO and HVO. There is ongoing debate within AVO and the VHP about the appropriate level of this sort of characterization. However, in order to fully understand and evaluate the risks from volcanoes to the flying public, better information about eruption frequencies and magnitudes is needed. Recent AVO-coordinated mapping campaigns at selected Alaskan volcanoes carried out by teams of USGS, other government, and university geoscientists have expanded the coverage of hazard assessment products. *The committee concludes that basic yet rapid assessment of the eruptive histories of as many of the Aleutians volcanoes as possible is necessary to guide prioritization of the placement of instruments used to provide warnings to pilots and other nearby infrastructure.*

Mapping Aleutian volcanoes has potential benefits beyond hazards assessment and mitigation. The Aleutians constitute one of the most active volcanic arcs on earth. Joint geophysical, geological, and oceanographic campaigns have recently been proposed to improve understanding of several different volcanic arcs. Although not the direct responsibility of AVO or the VHP, such studies could contribute significantly to the creation of a historical framework and better appreciation of eruptive activity in Alaska. Because the Aleutians are so active, they are important testing grounds for methods that will ultimately be applied in more populous areas.

Hawaiian Volcano Observatory

Although most visitors to Hawaii get the impression that its volcanoes erupt spectacularly but safely, HVO scientists have documented major volcanic hazards to both local and distant populations. The most common eruptive activity in Hawaii produces lava flows whose dangers are primarily to property (Figure 2.2(C)). Drifting clouds of noxious gas referred to as volcanic fog or "vog" represent an environmental health hazard in downwind areas.

Less common, but more dangerous, phenomena are also well represented in the geologic record. Violent explosive eruptions in 1924 and 1790 generated pyroclastic surges and showered Kilauea's summit area with large blocks. Recent studies have revealed many more such events than had previously been recognized. Sonar images of the seafloor adjacent to the Hawaiian Islands, collected and analyzed by VHP scientists in the 1980s, showed huge submarine landslide deposits that correlate with massive scars on the flanks of adjacent volcanoes. In at least some cases, these landslides must have been sudden and catastrophic, producing tsunamis that left marine deposits perched on the sides of nearby volcanoes, several hundred meters above sea level. These phenomena reflect a severe potential hazard for much of the Pacific Basin.

Faced with this array of dangerous processes and products, HVO has to employ a complex assessment strategy consisting of mapping, dating, modeling, remote sensing, and estimating probabilities. Most but not all of this evaluation has been targeted at the island of Hawaii (Appendix C), where volcanic activity is ongoing; and recent studies have been extended to Haleakala volcano on the island of Maui. Studies by HVO staff incorporating systematic mapping and radiometric age dating of lava flows and other deposits have created a comprehensive picture of the recent constructive history of the island. Based on this database,

HVO scientists have produced hazard zone maps that take into account the types of danger, the magnitude of typical events, and their frequency (Figure 2.3). These maps provide vital information, but the sharp boundaries separating hazards zones pose problems for civil defense authorities and insurance companies charged with interpreting their implications. Actual hazard potentials are smoothly varying continua and do not exhibit step functions like those portrayed by such maps.

During periods of sustained eruption, Kilauea emits about 2,000 tons of sulfur dioxide gas each day (Sutton, et al., 1997). This air pollution causes respiratory problems and contaminates rainwater-catchment systems that provide drinking water to many residents. HVO staff members closely monitor the amount and composition of gas emissions and collect and integrate information on volcanic air pollution from a variety of sources. They work closely with government officials and health professionals who inform residents and visitors about this hazard.

HVO faces additional assessment complications for those volcanoes that remain capable of erupting but have not been active in the past century (Hualalai, Haleakala, Mauna Kea). Recent mapping of these dormant volcanoes has clarified their hazards (Appendix C). Part of HVO's challenge is to convey this information to emergency managers and other public officials despite the widespread public perception that all volcanic activity is localized on Kilauea and Mauna Loa.

Cascades Volcano Observatory

The Cascades Volcano Observatory is responsible for assessing and monitoring the hazards of the volcanoes of the Cascades Range, which stretches from British Columbia to northern California (Figure 1.1). Two features of Cascades volcanoes most affect the assessment work of CVO: (1) many of them are located near major population centers; and (2) they typically lie dormant for decades, centuries, or millennia before returning to activity. These features mean that much of the work of CVO scientists involves public education about risks posed by seemingly benign mountains. Fortunately, widespread publicity about the Mount St. Helens eruption in 1980 raised general awareness about volcanic dangers. However, as time goes on with no nearby activity, this appreciation will fade. Because of the large number of volcanoes that could potentially erupt and the diversity of their eruption styles, CVO geologists have spent

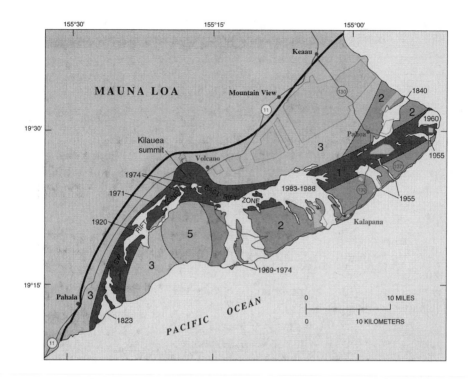

Figure 2.3 Hazards map for lava flows at Kilauea, Hawaii. Relative hazards range from 1 (high) to 5 (low). Lava flows erupted since 1823, gray; principal subdivisions, dark gray; boundary between Kilauea and Mauna Loa flows, heavy black line (USGS, 1992).

considerable effort mapping, dating, and collating information about the volcanic deposits of the region.

Cascades Range volcanoes have erupted about 50 times in the last 4,000 years, leaving substantial deposits. Seven of these eruptions took place in the past 200 years, and four affected areas far beyond the margins of the volcano. These frequencies suggest that there is approximately a 30 percent chance of an eruption in the region every 10 years and an 18 percent chance of an eruption whose influence extends beyond the base of the volcano.

Like other volcanic chains formed by the descent of one convergent tectonic plate beneath another, the Cascades have many different types of eruptions. Lava flows of highly varied chemistry, explosive blasts of ash and rock, massive flank collapses, and voluminous debris flows are all possible consequences when a Cascades volcano awakens. Several of the Cascades volcanoes have permanent glaciers, increasing the likelihood of dangerous debris flows. Even slight increases in hydrothermal activity near a volcano's summit may enhance glacial melting and debris flow formation. Thick ash deposits from explosive eruptions can also modify drainage patterns and choke streams and rivers, resulting in increased flooding even at great distances from the volcano. Although the 1980 eruption of Mount St. Helens had a serious economic and ecological impact on its surroundings, its scale is dwarfed by the magnitude of prehistoric eruptions in the region, such as those at Mount Mazama (Crater Lake) (Figure 1.1).

The two basic techniques used by CVO staff for assessing past Cascades activity—mapping and age dating—are much the same as those used in Hawaii. However, with a much larger number of volcanoes to evaluate, CVO staff members face more difficult prioritization issues than their colleagues at HVO.

Over the past three decades, successive groups of USGS scientists have been involved with volcano hazard assessment in the Cascades (Appendix C). Before the establishment of CVO, individual USGS geologists based mainly in Menlo Park took on long-term mapping projects of the major Cascades stratovolcanoes. These comprehensive studies were aimed as much at basic understanding of magmatic processes and regional geologic history as at determining the likelihood and distribution of future volcanic hazards. Many of these studies lasted for more than a decade, and they resulted in a large number of refereed scientific publications as well as geologic maps. A second group, based at USGS-Denver, rapidly generated more focused reports and hazards maps that outlined the dangers of individual volcanoes. In the 1990s, CVO scientists adopted a more collaborative approach, including sedimentologists and fluvial geomorphologists to identify a wider range of past activity. In addition, they placed greater emphasis on more rapid assessments, to ensure that all potentially active volcanoes have at least a basic hazards map that can be updated periodically when improved data and methodologies become available. CVO is also trying to evaluate the many smaller and less prominent volcanic centers between the major cones (Appendix C).

Long Valley Observatory

As a "virtual observatory" located in Menlo Park, California, and receiving monitoring data telemetered from Long Valley, the LVO has few resources available to devote to hazard assessment. However, Long Valley caldera and vicinity have been the subjects of several comprehensive mapping projects over the past 25 years. A priority for LVO staff has been to update these earlier reports and make them more consistent with assessments prepared by other parts of the VHP.

Despite its relatively restricted geography, the area monitored by LVO contains evidence of a diverse set of volcanic hazards. The greatest of these would be a caldera-scale explosive eruption like the one that formed the present physiography about 730,000 years ago. Neither scientists nor modern society has ever witnessed such a process, so recognizing its precursors is especially difficult. Seismic unrest and rapid uplift of the caldera floor in the 1980s caused widespread concern that a major magmatic explosion might be imminent. This uplift coincided with similar events at Yellowstone National Park; outside the city of Naples, Italy; and at Rabaul Volcano in Papua New Guinea. When the activity at the first two of these three sites subsided uneventfully, it highlighted the limitations of our knowledge of these most violent of eruptive phenomena. Assessment of the potential for this type of activity must rely exclusively on interpretations of mapped relationships caused by prehistoric events.

The most recent activity in the vicinity of Long Valley took place along the Inyo-Mono Crater chain within the past few hundred years. These events included the formation of several lava domes and explosive products. Geologic mapping suggests that the activity, fed by one or more dike-like intrusions, stretched for 11 and perhaps 25 km. Recognition in the late 1980s that potentially simultaneous eruptions could extend such great distances led to the construction of an escape road out of the town of Mammoth Lakes. Earlier mapping also revealed numerous young basaltic lava flows in the area.

Other Areas

The VHP may create other "virtual" observatories in the future in response to renewed activity at other volcanic centers. One such example is at Yellowstone National Park where, in the past several years, enhanced monitoring has been combined with earlier geologic and geophysical studies carried out by USGS and academic scientists. Bringing the state of assessment at this and other potential eruption sites around the western United States to a high standard requires access to as much compiled geological and geophysical information as possible. Thus, the scientist in charge of each observatory must strongly encourage all scientists to publish their results in a timely fashion.

Future of Hazard Assessment

The VHP's four volcano observatories have developed different approaches to hazard assessment that derive chiefly from differences in their histories and hazards. HVO is gaining excellent knowledge of the geologic history of those Hawaiian volcanoes most likely to endanger significant population centers. CVO is responsible for some volcanoes, such as Mount St. Helens and Mount Lassen, whose hazards are relatively well understood and others, such as Mount Baker and Mount Hood, for which documentation is much less complete (Appendix C). The volcanoes that AVO monitors are, in general, the least well known. These differences in baseline knowledge influence the types of data collection carried out by each observatory. HVO and LVO have fairly complete frameworks in which to insert newly mapped relationships. CVO oversees a mixture of both well-known and obscure volcanic centers. AVO is still in a mode of basic mapping and data collection.

If the VHP continues to be faced with a flat budget, it must find ways to carry out its mission more efficiently. **The committee recommends that the VHP initiate a form of collaborative prioritization with respect to hazard assessment** (Sidebar 2.1). This might include a broader application of the team approach now being used at AVO and CVO. It would require strong leadership from both VHP administrators and the scientist in charge of each observatory to ensure that program priorities are set and maintained and that the desires of individual scientists do not drive the program. In other disciplines and branches of the federal government, a demonstrated ability to prioritize is often rewarded with increased funding. Because volcanology is inherently multidisciplinary, volcano hazard assessment could provide the VHP and USGS with a flagship example of collaborative science.

The Decade Volcano Program, set up by the International Association of Volcanology and Chemistry of the Earth's Interior (IAVCEI) as part of the United Nation's International Decade of Natural Disaster Reduction, provided an instructive illustration of how collaborative prioritization can work in volcano hazard assessment. Fifteen volcanoes around the world were selected for comprehensive study on the basis of the size of the population at risk and the style of volcanic activity. Although the efficacy of the program was hampered by a lack of funding, in the most successful cases international teams of government and academic researchers carried out a coordinated regime of mapping, monitoring, and public education. Particularly noteworthy were the successes at Mt. Rainier, one of the "Decade Volcanoes," upon which scientists from the USGS; other federal, state, and county agencies; and a number of universities focused attention. Significant new insights were gained about the processes and dangers associated with these volcanoes. Importantly, the study of other volcanoes was postponed while these concerted campaigns were carried out. In some

---

SIDEBAR 2.1
Collaborative Prioritization

Collaborative prioritization occurs to varying degrees in many scientific disciplines. Astronomers, particle physicists, marine geoscientists, and planetary geologists have been forced into this mode by limited access to crucial facilities (telescopes, accelerators, research vessels, and spacecraft). Because of the large scale of these projects, overall programmatic funding levels tend to be higher, but grants to individual investigators may be smaller. In recent years, branches of science that have traditionally been less "high tech" have moved toward the collaborative approach. Field-based ecologists, geographers, and chemists have played key roles in the National Science Foundation (NSF) sponsored Long Term Ecological Research (LTER) program. Well-coordinated teams of life, earth, and physical scientists, in some cases supplemented by social scientists, study 22 LTER sites. Frequent meetings and widespread use of electronic communications ensure that individual projects are integrated and directed toward some common research goals. Another large NSF-supported program oversees the Science and Technology Centers, which are typically funded at $4 million to 5 million per year for up to 11 years. These cover a wide range of disciplines including earth science (e.g., Southern California Earthquake Center; Center for High Pressure Research).

cases, committees of scientists and public officials collectively decided what lines of research should be conducted and by whom.

In addition to prioritization, *volcano hazard assessment within the VHP would be improved by greater consistency of data collection, storage, presentation, and interpretation.* For instance, hazard maps are not prepared in a single easily understood format. Terms such as "high risk" and "low risk" mean different things to different scientists within the VHP and to members of the public. This lack of consistency also is found in VHP hazard maps. In general, communities accept hazards that could result in devastation of property and loss of life with annual probabilities of less than one in one million, but require mitigation strategies in cases where the probability is greater than one in one thousand. Between these values, decision making is more complicated and may involve questions about how much effort should be devoted to the mitigation. Some individual users may tolerate a completely different range of hazards. For example, a higher hazard level is acceptable for many structures and roads, such as those found in national parks, because such areas are easily evacuated during times of volcanic unrest. On the other hand, critical facilities, such as nuclear power plants, dams, and other large, expensive structures, generally require lower annual probabilities because damage to these facilities cannot be mitigated by evacuation and may greatly compound the disaster.

SUMMARY

Basic research in the VHP, although reasonably well integrated, is being threatened by budgetary and personnel constraints, that may diminish the program's ability to meet appropriate scientific goals. If these problems are not solved, the program will likely be forced to reduce levels of in-house basic research and/or to increase collaboration with non-USGS scientists. Hazard assessments, although traditionally strong in geologic mapping, radiometric age dating, and related activities, must be strengthened in modeling and probabilistic approaches, if the program is to continue to meet appropriate scientific goals. Existing hazard assessment activities at individual volcano observatories are effectively integrated and applied to hazard mitigation issues. The one-volcano, one-scientist projects currently under way, although scientifically appropriate, may not be effectively integrated with other studies or with the VHP as a whole.

3

Volcano Monitoring

WHAT IS VOLCANO MONITORING AND WHY SHOULD THE VHP DO IT?

Whereas volcano *research* seeks to explain all types of behavior of all volcanoes based on first principles and *assessment* aims to determine the long-range activity of a single volcano based on its past, *monitoring* looks at the short-term changes of a currently or recently active volcano in order to predict if and when a volcanic *crisis* might develop. To be effective, monitoring must be done before, during, and after eruptions and must be integrated with carefully designed communication schemes. It requires the type of long-term commitment of time and resources that academic and industry scientists generally cannot make. Furthermore, the quality of monitoring depends on the amount of experience of the participating scientists. For these reasons, the VHP is uniquely qualified within the United States to carry out volcano monitoring.

Monitoring strategies vary greatly depending on a number of factors such as the activity of the individual volcano, access, and available personnel and funding. Some volcanoes are monitored by a single seismometer, whereas others are covered by a comprehensive array of instruments. Certain monitoring methods, such as securing gas samples from fumaroles, require that scientists enter active vent areas. Other data can be collected remotely and with less risk, such as telemetered seismic and geodetic measurements or satellite-derived images or spectra. Volcano monitoring techniques can be simple (i.e., taking the pH of a thermal spring every several weeks) or complex (e.g., broadband source studies and seismic tomography).

Rapid advances in technology allow for more precise monitoring today than was imaginable when the VHP was formed. Monitoring of Kilauea was once carried out using manually-read water-tube tiltmeters and smoke drum seismic recorders, instruments that seem quaintly archaic today. At Mount St. Helens, deformation monitoring was conducted with electronic distance-measuring devices that determined changes in length with a precision of 20 mm over a 10-km-long baseline. Such measurements could be manpower intensive, but the data were relatively easy to reduce. Today, these measurements would likely be made with continuously recording global positioning system (GPS) receivers, supplemented by interferometric synthetic aperture radar (InSAR). GPS provides three-dimensional positioning with a precision of about 3 mm. InSAR can generate maps showing changes in vertical and horizontal position with a precision approaching 10 mm. Neither method requires personnel in the field (after initial installation); however subsequent data analysis can be highly complex. Similar changes are taking place in other subdisciplines, such as seismology and satellite-based remote sensing.

Done effectively, monitoring not only provides timely warnings to civil authorities of escalating hazards, but also leads to improved understanding and models of how volcanoes work. Monitoring generates baseline information against which changes in volcano behavior can be compared. These data are the essential ingredients with which scientific ideas and interpretations advance. Preserving the integrity and accessibility of data archives is thus essential if future volcanologists are to benefit from the decades-long records of volcano behavior gathered by the VHP.

A fundamental question facing the VHP is how to establish a balance between maintaining traditional methods that may be comparatively simple and inexpensive, and introducing new, more informative techniques that are more complex and costly. Archiving approaches have to be downward compatible, so that both old and new information may be accessed and compared in longitudinal or retrospective studies. Similarly, staffing decisions must ensure that sufficient knowledge is maintained about older data sets, even as younger scientists and technicians with new skills are recruited.

WHAT IS THE STATUS OF MONITORING WITHIN THE VHP?

The different VHP observatories monitor volcanoes in different ways. All observatories rely on seismic and geodetic instrumentation, such as tiltmeters, leveling, and GPS as their main monitoring tools. Other approaches are used selectively. For example, HVO employs a variety of instruments to monitor the gas flux from Kilauea and the amount of volcanic air pollution ("vog") in the surrounding area. CVO has developed innovative techniques for early warning of debris flows at Mount Rainier and other ice-clad Cascades volcanoes. AVO has focused on remote, near-real-time methods, mostly seismic and satellite based, to monitor volcanic unrest and eruptive activity at Alaskan volcanoes, since eruptions threaten the heavily used civil aviation routes traversing this region (see Sidebar 3.1 and Figure 3.1). LVO is making good use of various GPS techniques and strain meters for deformation and geodetic studies. The different techniques are discussed more fully below.

Overall, the VHP is doing an excellent job of monitoring volcanoes both within the United States and selectively in foreign countries under the auspices of the VDAP team. However, if one asks how the VHP's monitoring in 2000 compares with that 10 years ago, the answers are mixed. On the positive side, VHP staff have more experience today, and monitoring networks are more extensive and sophisticated. There has been a major expansion in Alaska as AVO has been developed of volcanic

---

SIDEBAR 3.1
Volcanic Ash Danger to Aircraft

Some of the world's busiest air traffic corridors pass over approximately 100 active volcanoes in the North Pacific capable of sudden, explosive eruptions. More than 10,000 passengers and millions of dollars in cargo fly across this region each day. On an average of four days per year, these volcanoes eject ash to altitudes of 30,000 feet, where most large jet aircraft fly. Since 1980, at least 15 aircraft have been damaged while flying through volcanic ash clouds. These clouds are difficult to distinguish from ordinary clouds, both visually and on airborne radar. Ash clouds can also drift great distances from their source. The particles from the June 15, 1991, eruption of Mount Pinatubo in the Philippines traveled more than 5,000 miles to the east and damaged more than 20 aircraft, most of which were flying more than 600 miles from the volcano. Ash clouds can diminish visibility, damage flight control systems, and cause engine failure.

On December 15, 1989, KLM flight 867, carrying 231 passengers bound for Anchorage, inadvertently entered an ash cloud from the erupting Redoubt volcano, 150 miles away. All four engines on the 747 failed when they ingested ash. The jet fell at a rate of 1,500 feet per minute from an altitude of 27,900 feet to 13,000 feet. After 11 tries, the pilot was able to restart the engines and land the plane safely. The incident occurred over the snow-covered Talkeetna Mountains, which have an elevation of 7,000 to 11,000 feet. The plane required $80 million in repairs, including replacement of all four engines.

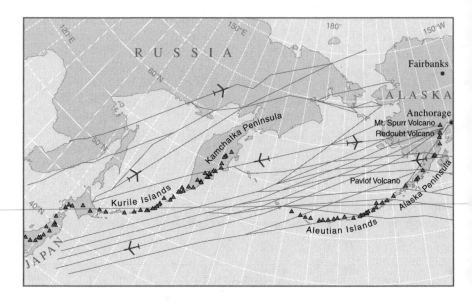

Figure 3.1 North Pacific and Russian Far East air routes pass over more than 100 potentially active volcanoes. Graphic designed by Sara Boore and Susan Mayfield.

plumes for aviation safety. There is better integration of diverse data sets, as growing numbers of VHP scientists appreciate the benefits of an interdisciplinary strategy. Finally, integrated studies of processes have provided a deeper theoretical understanding of how volcanoes work. On the negative side, the lack of new staff clearly has hindered monitoring efforts, particularly in the areas of physical volcanology, remote sensing, and gas studies. The VHP is also having a difficult time keeping up with technological advances in the core areas of deformation and seismic monitoring. Only a small number of VHP staff members know how to process GPS, InSAR, or broadband seismic data. As technology continues to improve, the VHP is in danger of being left behind; in the future it may not have the expertise to mount adequate monitoring campaigns. The VHP may want to initiate a retraining or continuing education program to allow scientists and technicians to expand their expertise.

Given this general state of affairs, one goal of this chapter is to identify weaknesses within the VHP monitoring strategy and propose needed changes. The committee believes that these problems must be addressed if the VHP is to remain capable of volcano monitoring. The status of different monitoring approaches is addressed first; then related issues, such as prioritization and access to wilderness areas, are examined.

Monitoring Approaches and Issues

Seismic and Deformation Monitoring

As magma approaches the earth's surface, volcanoes stretch and crack in characteristic ways. The combined seismic-deformation approach, which has traditionally been the core of VHP monitoring, tracks these phenomena to provide ample warnings of impending eruptions on most volcanoes. Seismic monitoring detects earthquakes that commonly serve as eruption precursors. Geodetic techniques reveal ground surface deformation associated with the movement of magma beneath volcanoes or with the development of flank instabilities.

More than half of the potentially active volcanoes in the United States have seismic stations, and instrumenting the rest is one of the stated goals of the VHP. The report *Priorities for the Volcano Hazards Program 1999-2003* (USGS, 1999) argues for an expansion of some existing networks and upgrading of overall instrument capability. The VHP also intends to improve established GPS networks that are measured sporadically with continuously recording arrays that enhance real-time forecasting. In addition, the use of permanent GPS receivers, borehole tiltmeters, and strain meters will be expanded. *The committee endorses these plans because they are directly applicable to the scientific goals of the VHP and will help to achieve hazard mitigation.*

Upgrading instrumentation is an ongoing challenge for any agency charged with monitoring natural phenomena. A balance is needed between preservation of traditional methods for which extensive data sets and staff expertise exist and newer approaches that require expensive retooling of instruments and personnel. The VHP faces such a quandary with regard to both seismic and geodetic data collection. Successful integration of, and migration to, new approaches will require the VHP to prioritize—scientifically, financially, and with respect to personnel. Difficult choices will probably have to be made, such as determining the optimal distribution of broadband seismometers. The VHP must proceed carefully and systematically so that older and newer methods are maintained simultaneously during the transition. Furthermore, VHP staff members must be trained in the new techniques. For example, volcano geodesy is a rapidly expanding discipline, with many promising new methods now coming on-line. The VHP has only a few participants in these efforts and has demonstrated relatively little administrative leadership in pushing for greater involvement. This appears to be a missed opportunity that could be rectified by additional hires, greater collaborations, and more widespread training programs.

Another issue discussed at length by the committee was real-time availability of monitored information, particularly seismic and geodetic data. Freely available data in a form that is easy to comprehend help educate the public, reduce suspicion during times of crisis, and allow non-VHP scientists with different perspectives and training to contribute to the interpretation of a volcano's status. On the other hand, unfettered access by improperly trained individuals to primary data from field instruments or remote sensing platforms could result in inaccurate interpretations, flawed policy decisions, and public panic. Although there are pros and cons for making data available on a real-time or near real-time basis, the committee believes that the advantages of public access outweigh the disadvantages. **The committee therefore recommends that VHP observatories take measures to make their data available on a near real-time basis.** A good example already exists at Long Valley caldera, where slightly processed USGS data have been accessible for several years on a near real-time basis as part of a World Wide Web site.

Finally, the committee was favorably impressed by AVO's attempts to install seismic networks (either large or small) on as many Aleutian volcanoes as possible. This is critically important work, since even a small eruption can significantly disrupt North Pacific air traffic. Instrumentation efforts should be piggybacked onto other activities (e.g., hazard assessments) wherever possible to maximize the monitoring capabilities within existing budgets. *The committee believes that a team approach for monitoring and studying Aleutian volcanoes from various perspectives should be expanded in the near future so that AVO can*

*provide airlines and other constituencies with adequate advance warning
of impending eruptions.*

Gas Monitoring

The collection of volcanic gas data is another essential monitoring
tool that complements seismic and geodetic information. Changes in the
permeability and fracture system of a volcano may be reflected in gas
discharges before seismic or deformation instruments record magma
ascent through newly established pathways (e.g., during the 1994
eruption of Popocatepetl, Mexico). The committee was disturbed to learn
of the paucity of gas geochemical expertise and utilization within the
VHP. *The program should reestablish in-house capacity to use and
develop both conventional and novel methods for measuring and
interpreting volcanic gases.* In addition, AVO has to expand its gas
geochemical capabilities, rather than relying totally upon personnel at
CVO. The committee notes that similar recommendations were made
during an internal review of the VHP in 1986 (Shoemaker et al., 1986).

One of the priorities of the VHP is to support efforts to improve
airborne and continuous ground-based techniques for quantifying gas
emissions at restless volcanoes. The program also plans to better
integrate seismic and geodetic monitoring with gas measurements. The
committee encourages the VHP to develop and install in situ gas monitoring devices on many more volcanoes so that real-time geochemical
data can be collected in conjunction with seismic and geodetic information. Continuous CO_2 monitoring is currently being carried out on
Mammoth Mountain at Long Valley caldera; however, the conditions
there are much less harsh than in an active crater where acid gases cause
corrosion and interference. The only active crater in which the VHP has
routinely deployed in situ gas detectors is Kilauea, but even there,
monitoring of SO_2, CO_2, HCl, and HF is uncommon.

New ground-based instruments for remote sensing of CO_2 and other
gases are currently being developed outside the USGS, including Fourier
transform infrared spectroscopy (FTIR), gas correlation spectroscopy
(e.g., GASPEC, MicroMaps), and light intensity detection and ranging
(LIDAR) techniques. These instruments have major technical advantages
over existing approaches used by the VHP. For instance, they can rapidly
measure gas ratios that reflect the degassing state of magma and its

eruption potential. *The committee believes that VHP scientists should be in the forefront of such efforts, either by obtaining this equipment themselves or by actively collaborating with groups who are developing these tools.* Furthermore, to better understand the context in which gas readings are obtained, VHP gas personnel have to interact closely with researchers who will be using Earth Observing System (EOS) instruments (e.g., ASTER) to measure SO_2 emissions from volcanoes into the troposphere.

Hydrologic Monitoring

Although less prominent in the public's awareness than lava flows or pyroclastic phenomena, mixtures of volcanic debris and water are among the most deadly products of volcanoes. Detection of volcanic debris flows close to their sources can provide timely warnings to people in downstream areas. Many of the casualties from the Mount Pinatubo eruption were caused by debris flows that developed weeks, months, and years after the magmatic output ceased and tens of kilometers from the vent. Automated detection systems developed in the United States were installed by VDAP staff members and their Filipino collaborators, providing greater lead times for evacuations.

Several types of electronic instruments can detect and monitor debris flows at active volcanoes, but none are currently foolproof. For instance, trip wires are difficult to install and are subject to vandalism and accidental breakage. Conventional seismographs can note the passage of a debris flow but cannot pinpoint its location, nor can they distinguish such a flow from other sources of persistent noise such as rain or wind.

Scientists at CVO are currently developing two automated systems for detecting debris avalanches and debris flows. In the first, acoustic flow monitors (AFM) sense and analyze ground vibrations with a compact, solar-powered unit that is installed in specific stream channels. This can provide up to several hours advance notice of an approaching debris flow. AFM development is a prime example of how theoretical and laboratory studies can have direct bearing on monitoring capabilities because the results of flume tests and physical process models were used to refine early versions of the instruments. A second approach uses single-channel seismic sensors, pressure-transient counters, and/or lightning detectors transmitting data via low-baud-rate Geostationary

Operational Environmental Satellites (GOES). This system identifies the onset of activity in places where network coverage or visual confirmation is limited or impractical.

Over the next five years, the VHP plans to improve and field-test remote eruption detection stations for possible deployment in the western Aleutians and the Cascades. *The committee supports this goal because it is relevant to the VHP mission to mitigate volcano hazards.* The VHP should also explore ways to better monitor groundwater flow and pore pressures within volcanic edifices. This type of information could help establish the potential for phreatic and phreatomagmatic activity, sector collapse, and internal pressure buildups capable of generating explosive blasts. *Such hydrologic monitoring warrants greater attention by the VHP.* The incorporation of glacier budget studies as part of VHP monitoring on ice-clad volcanoes would also contribute to this goal.

Satellite Remote Sensing

Another VHP goal that the committee fully supports is the continued development of near real-time remote sensing of volcanoes and their associated ash clouds in areas that are difficult to access. Most of the VHP's remote sensing work is centered at AVO, where satellite data are used to identify thermal anomalies and track eruption plumes and where inclement weather makes traditional observations of volcanoes more difficult. Remote sensing data are only slowly becoming integrated into the monitoring strategies of the other VHP observatories, primarily because of a lack of resident expertise.

A new generation of EOS instruments is now providing potentially useful information for volcano monitoring (e.g., data on thermal regimes, SO_2 gas emissions, deformation, and digital topography). *The committee believes strongly that the VHP should take advantage of this opportunity to the fullest extent possible,* through two steps: (1) Expand the remote sensing program at AVO and officially designate it as the center of internal expertise for the VHP. In this manner, other observatories will gain access to these tools without having to hire as many of their own specialists. A secondary benefit will be to reduce the sense of isolation felt by AVO personnel relative to the rest of the VHP, especially if training sessions for mainland and Hawaiian colleagues are held at AVO. (2) Forge stronger links with universities and government laboratories where such expertise is available and where many new techniques are being developed. In particular, **the committee urges the USGS to work with NASA to argue in support of an InSAR satellite specifically designed for hazards monitoring** (Sidebar 3.2). However, the major funding needed for development and deployment of such a satellite cannot come out of the VHP budget. Other federal agencies are also striving to expand their use of remote sensing capabilities, and this offers additional partnering opportunities to the VHP. For instance, the U.S. military became much more sensitive to volcanic dangers during the 1991 eruption of Mount Pinatubo. The Navy's Pacific fleet (under supervision of CINCPAC (Commander-in-Chief, U.S. Pacific Command)) now monitors the status of all types of natural disasters around the Pacific Rim, in part through the efforts of the Pacific Disaster Center (PDC). The PDC is a federally funded facility with the mandate to provide disaster managers with value-added information on all types of natural and human-made disasters. Although it has excellent remote sensing and geographic information system (GIS) capabilities, the PDC has limited volcanic expertise in terms of monitoring and numerically modeling the many different types of eruptions that create hazards around the Pacific and Indian Oceans. Encouraging more coordination between the remote sensing activities of the PDC and volcanic monitoring by HVO, AVO, and CVO would likely benefit both groups. Other agencies that could gain from an enhanced VHP remote sensing presence include Federal Emergency Management Act (FEMA) and the FAA. Another potentially promising remote sensing partner for the VHP is the Global Disaster Information Network (GDIN), which is expected to evolve a global perspective for natural hazard monitoring comparable to the PDC's role around the Pacific and Indian Oceans.

The committee also considered the potential value to volcano monitoring of two existing remote sensing programs based outside the VHP, the Hazard Support System (HSS) and the Center for Integration of Natural Disaster Information (CINDI). The HSS is designed to use classified military "spy" observations for civilian purposes. Beginning in FY 2000, this program will be partially funded by the National Mapping Division of the USGS. A single senior VHP member in Reston, Virginia serves as liaison to HSS, helping to find ways to use these data for volcano monitoring. CINDI is an unclassified USGS facility that develops and evaluates technology for information integration and dissemination.

SIDEBAR 3.2
Use of InSAR to Monitor Volcanic Deformation

Yellowstone National Park holds a very large and restless caldera, which formed approximately 630,000 years ago through the ejection of about 1,000 km^3 of debris. Explosive activity was followed by rhyolitic lava flows extruded from 150,000 to 70,000 years ago. Although Yellowstone caldera has remained dormant since then, geological and geophysical evidence suggests that an underlying crustal magma reservoir remains partially molten because of periodic intrusions of basalt. Since another caldera-forming eruption is possible, a continuous monitoring program is maintained by the VHP in conjunction with academic scientists.

A variety of geological and geophysical field studies in the past 20 years have revealed that Yellowstone has experienced uplift and subsidence of its caldera floor in historic and prehistoric times. Recently, InSAR data from the European Space Agency have been used to dramatically pinpoint the deformation of the entire caldera floor (Wicks et. al., 1998). The interferogram on the left, made from data collected between August 1992 and June 1995, shows about 60 mm of subsidence. The one on the right, based on data from July 1995 and June 1997, indicates 30 mm of uplift.

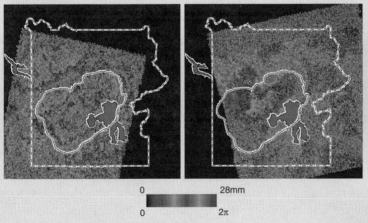

The great promise of InSAR is that it requires no ground-based presence. Thus, precise monitoring can be effected without placing scientists or technicians in harm's way. Furthermore, there is no need to establish ahead of time which volcanoes are going to become active. InSAR potentially allows practically all volcanoes to be monitored. However, none of the existing or planned SAR systems is optimally designed for volcano monitoring.

CINDI is also involved in research in data integration, analysis, and modeling and provides ongoing support of the evolution of the USGS processing and delivery of hazards data.

Although these programs have clear promise for monitoring and have been strongly promoted by some non-VHP members of the USGS, the committee heard two primary reservations expressed by several VHP scientists. The classified nature of some of the data and the fact that military priorities control which observations are made mean that VHP personnel may have limited access, controlled by the Department of Defense. This adds an extra bureaucratic layer of communication and interpretation, slowing responsiveness and potentially reducing the effectiveness of the monitoring effort. Second, because these programs are very expensive, they run the risk of draining sparse resources away from the VHP for questionable returns. For these reasons, *the committee cautions against greater involvement with CINDI and HSS unless and until better assurances can be obtained about data access and cost containment. A potentially less problematic alternative would be to establish closer ties with the nonclassified EOS program run by NASA.* To best incorporate satellite- and aircraft-based remote sensing programs such as EOS into its monitoring mission, the VHP should expand partnerships with academia, other government agencies (e.g., NASA, NOAA), and private sector groups. Caution should be exercised in placing too much reliance on classified remote sensing data.

Other Monitoring Issues

Prioritization of Monitoring Activities

The scientific value of basic monitoring is often underappreciated, yet the resulting data provide the framework with which new ideas are developed and tested, models constructed, and forecasts made. Just as with hazard assessment, the type of monitoring that is done by the VHP at a particular observatory depends in part on the experience and biases of the people who work there. A less ad hoc approach would include objective, program-wide evaluation of new techniques followed by observatory-wide decision making by the scientists in charge or by a committee of users. Any new techniques for data evaluation must be downward compatible, so that older data sets remain available for future studies. Better communication among the different scientists in charge or other representatives helps ensure that all observatories learn of the latest monitoring technologies in a timely fashion. This practice would be more "top-down" than current procedures and would potentially reduce the autonomy of individual scientists and observatories.

Prioritization across government agencies would also help the VHP with its monitoring mission. Regular interactions between the scientists in charge and representatives from other science-oriented agencies, such as NASA, NOAA, DOE, FEMA, and the NSF, would allow for coordination of expenditures and extramural grant funding related to monitoring activities. These meetings would also let the scientists in charge notify their counterparts about the types of information the VHP needs to improve monitoring for public safety. For instance, if U.S. Forest Service and National Park Service managers were included in these discussions, they might better appreciate why VHP scientists need occasional vehicular access to otherwise off-limits wilderness areas (see next section).

Access to Wilderness Areas

Many volcanoes in the Cascades and several in Alaska lie within wilderness areas and other lands managed by the National Park Service and the U.S. Forest Service. This situation creates a conflict between the need for effective monitoring in order to serve public interests and the desire to minimize mechanized access to the areas in question. The use of four-wheel-drive vehicles or helicopters to install and maintain monitoring equipment has been greatly restricted under interpretations of the Wilderness Act. Recent events have led members of the scientific community to assert that "there is no clear policy on research in parks and wilderness" (Eichelberger and Sattler, 1994; Eichelberger, 1997). Clearly, this is a delicate situation that must be resolved by effective communication among the different agencies involved. The problem requires that scientists in charge of the observatories maintain a close working relationship with their counterparts in nearby National Parks and National Forests. Hawaii Volcanoes National Park provides a useful example of a flexible policy, which in part grew out of recognition by the communities situated near active volcanoes of the importance of good access for monitoring. *High-level administrators within the USGS and other organizations must actively campaign to gain recognition that monitoring efforts require special attention and priority.*

SUMMARY

Persistent budget problems place four types of constraints on the VHP's ability to monitor volcanoes. (1) Aging equipment is not replaced soon enough (or at all), which increases the chances of failure during a crisis. (2) The VHP's traditional role as the developer and tester of new monitoring equipment and techniques is jeopardized. (3) The number and extent of regular instrumented surveys, which are crucial for the success of any monitoring program, are restricted. (4) Personnel familiar with new techniques are not hired. If the current situation is not reversed, the VHP may not be able to field the best instruments or maintain its traditional high standards for monitoring. These issues apply in varying degrees to all of the monitoring methods used by the VHP, and if they are not addressed in the near future, the program runs the risk of being unable to meet appropriate scientific goals. On the other hand, the monitoring methods currently employed in the VHP seem to be well integrated and applied to achieve hazards mitigation.

4

Crisis Response and Outreach

WHAT IS VOLCANO CRISIS RESPONSE AND WHY SHOULD THE VHP DO IT?

The third operational component of the VHP mission, after *assessment* and *monitoring*, is *crisis response*. The transition from monitored volcanic activity to a volcanic crisis has as much to do with potential societal impact as with the nature of the eruptive phenomena. For example, mitigation of the nearly 20-year eruption of Kilauea volcano that began in 1983 passed back and forth between routine monitoring, when no human infrastructure was immediately threatened, to crisis response, when subdivisions and highways lay in the path of advancing lava flows. Similarly, the Mammoth Lakes resort area of eastern California had a serious volcanic crisis in the early 1980s triggered by a large number of earthquakes, even though no magma broke the surface. In contrast to research, assessment, and monitoring, which all can be carried out largely by scientists and technicians with minimal involvement of the general public, crisis response requires close coordination with civil defense officials and the potentially affected population. Consequently, this aspect of the VHP requires the integration of a broader set of political and social science skills than an organization of geoscientists would normally be expected to possess.

In the United States, the USGS is expressly and uniquely empowered by the Stafford Act (Public Law 93-288) to issue timely warning of potential volcanic disasters to affected communities and civil authorities. Warnings can lead emergency management officials to evacuate people and property from areas of high hazard and can help educate the public about the possible impacts of impending eruptions. The VHP maintains the capability and protocols for rapidly deploying response-ready staff and monitoring equipment. Since 1980, the program has provided major response to 10 domestic volcano-related emergencies: Augustine, Redoubt, Spurr, Akutan, and Pavlov volcanoes in Alaska; Long Valley in

California; Kilauea, Mauna Loa, and Loihi in Hawaii; and Mount St. Helens in Washington (Figure 1.1).

Although not an explicitly mandated part of its mission, the VHP has also developed an international crisis response capability, the Volcano Disaster Assistance Program. Following the disastrous eruption of Nevado del Ruiz, Colombia, in 1985, which killed more than 23,000 people, the USGS and the U.S. Office of Foreign Disaster Assistance (OFDA) developed VDAP, headquartered at CVO, to respond to select volcanic crises around the world. VDAP has proven to be highly effective in saving lives and property by assisting local scientists in determining the nature and possible consequences of volcanic unrest and communicating eruption forecasts and hazard mitigation information to local authorities. VDAP has responded to 15 international volcanic crises since 1980 (Figure 4.1).

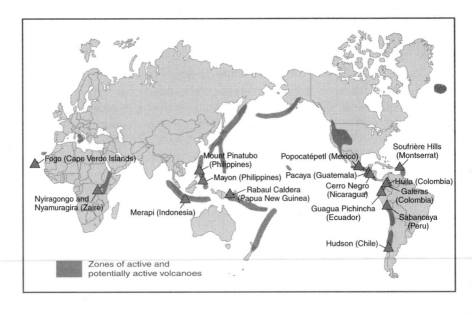

Figure 4.1 Map showing volcanoes to which VDAP has been deployed since the program was formed in 1986. Graphic designed by Sara Boore and Susan Mayfield.

VDAP offers direct and indirect benefits to the VHP and the U.S. government. Foreign responses provide valuable training for scientists from both the USGS and the affected country. Foreign responses also offer the VHP a statesmanship role and allow for the development of expertise and institutional capabilities in other countries so that they can better deal with subsequent crises both in the United States and abroad. They allow more frequent testing of equipment and techniques, eruption models, and process theories than would be possible solely from domestic responses. VDAP also helps further the global interests of the U.S. government by protecting U.S. businesses and military installations in foreign countries where large volcanic eruptions occur. With the growing globalization of the economy, damage to U.S. companies and industries from natural disasters in any part of the world can have both immediate and long-term impacts on domestic economic security.

WHAT IS THE STATUS OF CRISIS RESPONSE WITHIN THE VHP?

A successful volcano crisis response is built on information gathered during process-oriented research, hazard assessments, and monitoring campaigns. Fundamental research helps refine eruption models so that they can be applied more accurately to real situations. Prior hazard assessments provide essential clues about the possible impacts of future eruptions. A comparison with baseline monitoring data collected over years or decades gives the response team clues about the rate of development of the crisis.

Once unrest has been detected, a much more extensive instrumentation suite may be deployed on and around the volcano, and more personnel can be brought in to assist with data interpretation. If significant danger is thought to exist, a team of scientists from both inside and outside the VHP is assembled. It is essential that this group be able to work collaboratively, under increased public scrutiny, for long hours, and under intense pressure. It must be able to evaluate and update existing emergency response plans and to communicate them effectively to civil defense officials and the media.

Each VHP observatory has its own crisis response needs and protocols. At HVO, the eruption that began on Kilauea's east rift zone in 1983 continues unabated to this day; and this has led to an ongoing atmosphere of alert, sometimes punctuated by true crisis, that has strained the observatory's human and instrumental resources. The future response of HVO to this eruption might well be reassessed so that staff members could be freed for other tasks. AVO's responses are directed chiefly at the needs of the aviation community: rapid and accurate confirmation of the existence of ash plumes followed by up-to-date trajectory information. CVO has not had to deal with a local crisis since the end of the eruption of Mount St. Helens nearly 15 years ago, although CVO scientists have responded to crises elsewhere, including Long Valley.

When the VHP is asked to respond to an international crisis, it uses a systematic approach. At the request of host countries and in conjunction with the USAID, an experienced team of USGS and other scientists can be dispatched rapidly to developing volcanic crises with a portable cache of state-of-the-art monitoring equipment. In contrast to domestic crisis response, foreign deployments generally build upon monitoring and assessment efforts carried out by agencies of other governments following protocols that may be significantly different from those used by the USGS.

The USGS contribution to VDAP includes two seismologists, several engineers and technicians, administrative and outreach specialists, links to the academic community, and access to volcanologic expertise throughout the agency. The USAID provides half-time salary support for seven core scientists, a full-time geoscience adviser to the OFDA, and an emergency fund that covers rapid response, technical assistance visits, collaborative scientific work, training for staff in developing countries, an equipment cache, and development funds. Other beneficiaries of VDAP, such as the overseas offices of U.S. corporations or the U.S. military, make very limited or no financial contributions to sustaining the program.

The committee heard widespread praise for the scientific quality and commitment of the VDAP team. This group has a proven track record of directly and sensitively assisting many countries. Perhaps best documented is the response to the eruption of Mount Pinatubo, Philippines (see Sidebar 4.1), during which accurate prediction of the timing and magnitude of the explosive phase led to great savings of lives and property. The VDAP response left the PHIVOLCS better able to deal with future volcanic crises.

SIDEBAR 4.1
Response to Mount Pinatubo, Philippines

On April 2, 1991, after being dormant for 500 years, Mount Pinatubo in the Philippines awoke with a series of steam explosions and earthquakes. About 1,000,000 people lived in the region around the volcano, including about 20,000 American military personnel and their dependents at the two largest U.S. military bases in the Philippines: Clark Air Base and Subic Bay Naval Station. The slopes of the volcano and the adjacent hills and valleys were home to thousands of villagers. VDAP responded immediately by joining the Philippine Institute of Volcanology and Seismology and installing monitoring instruments and interpreting deposits from previous eruptions.

On the morning of June 15, 1991, Pinatubo experienced the largest volcanic eruption on earth in more than 75 years. The most powerful phase of this eruption lasted more than 10 hours, creating a cloud of volcanic ash that rose as high as 35 km and grew to a diameter of nearly 500 km. Falling ash covered an area of thousands of square kilometers, and pyroclastic flows filled valleys with deposits of ash as much as 200 meters thick.

Estimates show that the monitoring performed by the USGS and PHIVOLCS saved at least 5,000 lives. The alerts allowed residents living around Pinatubo to flee to safety. It also permitted more than 15,000 American military personnel and their families to evacuate Clark Air Base for safe locations. In addition, property worth hundreds of millions of dollars was protected from damage or destruction in the eruption. Aircraft and other equipment at the U.S. bases were flown to safe areas or covered, and losses of at least $200 million to $275 million were avoided. Philippine and other commercial airlines prevented at least another $50 million to 100 million in damage to aircraft by taking similar actions. Other commercial savings and the sentimental or monetary value of the personal property salvaged by families are difficult to quantify but nonetheless important. These savings in lives and property were the result of quick response and monitoring of Mount Pinatubo by scientists in PHIVOLCS and VDAP.

WHAT ARE THE OBSTACLES TO SUCCESSFUL CRISIS RESPONSE BY THE VHP?

Although crisis response is clearly one of the most successful aspects of the Volcano Hazards Program, the committee heard several suggestions for ways in which these functions could be improved. These ideas, which parallel proposals raised in Chapters 2 and 3, fall under the headings of training and knowledge dissemination, infrastructure and budget, and partnerships. Here the committee highlights some of these potential problem areas and makes recommendations for their solution. Additional suggestions can be found in Chapter 5.

Training and Knowledge Dissemination

Often the most valuable asset for a scientist responding to a crisis is relevant prior experience. Because its members are exposed to a wide variety of eruption styles and settings, VDAP offers the most effective way to prepare VHP staff for future domestic crises. The present system for selecting non-VDAP members of the VHP to join its foreign deployments appears too haphazard. **The VHP should implement a more formal mechanism for participation in VDAP to see that as many people as possible are exposed to this type of training.** In-service workshops provided to staff members at VHP observatories by VDAP personnel would be another way to disseminate the knowledge gained at foreign volcanoes. These workshops, by allowing a number of staff members from various observatories to discuss crisis situations, provide good opportunities for team building prior to the onset of an actual eruptive crisis to which they must respond.

Another missed opportunity for expanding the training potential of foreign volcanic crisis responses comes from the inability of VDAP members to archive their observations. In the high-stress environment of a volcanic eruption, scientists and technicians rarely have the time to provide good documentation (to national standards where they exist), archiving, and open access to their data. Yet these data could be used for background studies in preparation for dealing with new unrest elsewhere in the world. A possible solution would be to create one or more documentation and archive specialist positions for each response team, somewhat like the journalists assigned to military units during wartime. *The success of VDAP should be measured not only in terms of mitigation of eruption impact, but also in terms of how well information and knowledge are disseminated in anticipation of future crises.* This change of strategy might ensure greater access to data that could be used to prepare future crisis teams.

A related programmatic issue is how staff members balance their responsibilities. Even if assistance were provided for archiving and distributing data from volcanic crises, individual scientists still have to incorporate their experiences into the published scientific literature. There is no question that the VHP and USGS benefit from the firsthand, frontline knowledge gained by VDAP workers in crisis situations. However, timely publication of the results enhances the flow of information that eventually leads to better understanding, improved forecasting, and crisis response. In addition, by publishing their observations, techniques, and conclusions, the scientific staffs maintain their stature and credibility. This issue demands close monitoring, coordination, and allocation of staff time by the relevant scientists in charge to ensure that such information is forthcoming. *The stated VHP goal of carefully documenting actual volcanic crises and responses is extremely important if the maximum information is to be obtained from any given eruption and is strongly endorsed by the committee.*

Infrastructure and Budget

In addition to the valuable staff training opportunities provided by VDAP missions, foreign responses also allow new hardware and software to be evaluated under crisis conditions. Technical development of new instrumentation requires field tests for accurate calibration. Domestic volcanic crises are generally too infrequent (Cascades), too inaccessible (Alaska), or too limited in scope and applicability (Hawaii) to permit adequate assessment of a broad range of equipment needed during a serious eruption. In contrast, foreign volcanic activity occurs in a wide variety of geologic, climatic, and sociopolitical settings, allowing for much more rapid calibration and improvement of new instruments.

A consequence of continuing tight VHP budgets has been the growing obsolescence of much of the equipment used in crisis response. One way the VHP can extend its equipment budget is to partner with manufacturers and other government agencies that design new instruments. In the 1970s and 1980s, the national laboratories of the Department of Energy had programs to develop new techniques for monitoring volcanoes. In the past decade, NASA has supported a working group focused on finding better ways to monitor active volcanoes using remote sensing. The Department of Defense is actively involved in the development of microelectronic sensors of various types that can operate in extreme conditions such as volcanic craters, flows and plumes. Coordination between the VHP and programs such as these could help stretch the limited funds available for crisis response while expanding the range of information obtainable from dangerous volcanoes. **The committee encourages the VHP and VDAP to work more closely with NASA, DOE, DOD, and NOAA, as well as with NSF-funded consortia like University NAVSTAR Consortium (UNAVCO) and Incorporated Research Institutions for Seismology (IRIS), in the development of new instrumentation and approaches suitable for detecting the conditions within erupting volcanoes.** However, the testing of new instruments and methods should never compromise VDAP's ability to effectively respond to a volcanic crisis.

The current level of VDAP funding allows a maximum of one deployment at a time, leading to occasional difficult decisions about priorities when multiple crises occur almost simultaneously. Available funding also does not allow the VHP to hire replacements for those VDAP scientists who are prevented from carrying out their domestic assignments because of extended foreign commitments. Furthermore, budget constraints limit the number of scientists able to participate in crisis response activities. This in turn reduces the ability of the deployments to serve as training opportunities for workers responsible for domestic crisis response. *The committee unconditionally supports the stated VHP desire to expand the size of the VDAP.*

Partnership Issues

Crisis responses provide opportunities to better link the VHP with outside groups such as university faculty, foreign scientists, and emergency management officials. In the past, university researchers have commonly been excluded from the volcanic crisis situations coordinated by the VHP. Strengthening ties with faculty members and their students by integrating them into VDAP teams could provide long-term benefits to the VHP, by bringing in new perspectives and expertise, and to the greater volcanological community, by helping to train the next generation of volcanologists. There must, however, be clear ground rules for such participation, especially in international responses, because non-VHP personnel are not directly answerable to the local government. In particular, academic scientists should avoid increasing local confusion by offering opinions to the press or to civic officials that conflict with those expressed by the coordinating government agency. The involvement of non-VHP personnel should contribute to the quality of the crisis response and to the VHP's stated goal of improving forecasting ability. Students could be responsible for ongoing routine measurements and could provide help with data archiving, thus enabling the professional team to spend more time on data integration and assimilation. International responses should be coordinated with local investigators; VHP members have generally accomplished this and have developed professional relationships in other countries on the level of both individual scientists and institutions. One goal of this collaborative work should be greater use by all parties of deterministic forecasts based on theoretical models rather than more widespread, purely empirical approaches.

Establishing working relationships with local emergency management officials before the onset of a crisis is an important, though difficult, goal. The formulation of local emergency plans is an excellent way to set up these relationships before they are put under pressure in a crisis. Though the VHP has already been done much work in this area and good collaboration exists, this should be enhanced and expanded. As in all partnerships, the roles and responsibilities of all parties have to be defined clearly.

Another way to partner during volcanic crises is to use "expert elicitation" (Sidebar 4.2), a technique that relies on group expertise to evaluate and prioritize different scenarios when available data are inherently ambiguous. Recently, this approach was used to assess the likely eruptive behavior of the Soufriere Hills volcano on the island of Montserrat. A globally distributed group of experts set up a decision tree to attempt to forecast future volcanic behavior. Probabilities of specific eruptive outcome "branches" (timing, magnitude, products) on this tree were estimated by the collective best judgment of the participants. The group was re-polled periodically to allow members to debate, incorporate, and respond to the latest events. In this way an assessment could evolve continuously, in parallel with changing eruptive conditions. It allows individuals to focus on those factors that are the most uncertain. This is a promising approach that has to be refined through wider application, both inside and outside the VHP.

SIDEBAR 4.2
Expert Elicitation

Expert elicitation is a useful technique when data are open to alternative interpretations, yet decisions based on these data must be made. In some implementations of expert elicitation, such as during the extended volcanic crisis at Soufriere Hills volcano, Montserrat, a hazard event tree is constructed and transition probabilities are estimated by a team familiar with the volcano and likely eruption outcomes. Event trees, which may vary from volcano to volcano, graphically illustrate the potential outcomes of volcanic unrest and eruption in a clear sequence.

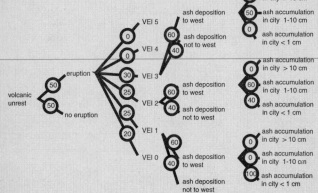

The likelihood of activity progressing to a given state is governed by transition probabilities at each branch of the hazard event tree. Each of these transition probabilities is debated and estimated by the expert team.

In an expert elicitation during volcanic unrest, each expert volcanologist would have to create such a logic tree and defend the branches and the transition probability that she or he estimates for each possible outcome. Ideally, this would make the reasons for variation in the estimated transition probability clear.

The advantages of using event and logic trees during episodes of volcanic activity include the following:

- The progress of volcanic activity and possible outcomes are clear to everyone involved.
- Discussion among scientists can focus on tractable questions.
- Differences in scientific opinion are identified, and therefore are more easily discussed.
- Widely varying interpretations can be weighted.
- A simple record of decision making is produced.

A point to remember is that consensus is not always correct. Expert elicitation is not a substitute for data, and the elicitation is most valuable when it is treated as a very dynamic process in which probability distributions change continuously as the analysis proceeds. Provided this is understood, expert elicitation can be a useful tool to explore variation in likely outcomes and to understand the origins of uncertainty in eruption forecasts.

HOW DOES CRISIS RESPONSE RELATE TO PUBLIC OUTREACH?

Just as the *acquisition* of volcanic knowledge takes place within a continuum between process-related research and hazards assessment, so does the *application* of this knowledge occur within the spectrum between crisis response and public outreach. The USGS in general and VHP workers in particular see their primary role as being advisory to civic officials responsible for making decisions about public welfare, rather than devising such policy themselves. This distinction leads to some potentially awkward situations, since the communities surrounding an active volcano have great interest in the information obtained by VHP geoscientists and apply pressure to politicians to make it available. The VHP must remain sensitive and responsive to the public's desire for interpreted hazard data, especially during a crisis. At the same time it has to avoid creating panic by releasing premature or inadequate information about ongoing or imminent eruptions.

Recent improvements in mapping and monitoring techniques and expansion of the Internet and telecommunications infrastructure mean that opportunities for the VHP to communicate with the public are greater than ever. Thus, in the United States, eruptions are less likely to have unexpected and disastrous consequences than in the past. On the other hand, as populations, economic development, and tourism increase in volcanically active regions, the VHP will face greater pressure to continuously update and improve existing hazard assessments and to keep the public informed about the status of nearby volcanoes.

In a crisis situation, the scientist in charge of the relevant observatory plays a crucial communication role. Although the USGS is not responsible for acting on the hazard assessments it produces, once a crisis develops the scientist in charge can and should provide clear evaluations of the evolving situation to all involved public officials. To the extent possible, the scientist in charge must have already established trust among members of other government agencies and the public so that their recommendations will be taken seriously.

In times between volcanic crises, communication and educational outreach remain essential functions of the VHP. Of the many forms of outreach, perhaps the most important is establishing good relations with local civil authorities, especially those concerned with emergency management. Many of these interactions require one-on-one briefings, but others consist of lectures, workshops, and field excursions during which dialogue is established with larger groups of officials. These individuals, once made aware of the hazards posed by volcanoes, play critical roles in disseminating this information to the concerned public and generating support for other VHP activities. Many VHP staff members, including the scientists in charge, have successfully carried out such communication.

The VHP also communicates directly with the broader public in many ways. These include one-page fact sheets, simple maps and brochures, videos, exhibits, booklets, and presentations to citizen groups. In recent years, large amounts of information have been made available through observatory Web sites. Members of the HVO staff write a weekly article entitled "Volcano Watch" that is published in a local newspaper, and a group of volcano specialists from the CVO has set up a "volcano hazards booth" at the Western Washington State Fair, downslope from nearby Mount Rainier. Collectively, these products and activities reach a broad cross section of the public and do a good job of informing them about volcanoes and associated hazards.

What Are Some Obstacles to Successful Outreach?

Although the committee finds the VHP's varied outreach activities to be highly successful, there are some steps that the VHP, the USGS in general, and DOI could take to improve these programs. Many VHP outreach products, such as maps, booklets, and other "hard-copy" items, are sold at museums and other public institutions. However, either by law or by USGS policy, the program cannot retain proceeds from these sales. This is unfortunate, because the income could be reinvested, providing both the incentive and the means to generate new outreach products or to disseminate existing ones more widely. The National Park Service (NPS) has addressed this situation by fostering "natural history associations" to promote the interests of many of its parks. These associations, outside the official NPS structure but located within park boundaries, operate much like usual nonprofit enterprises and are able to retain proceeds from the items they sell. A USGS "Volcano Hazards Association" might operate in much the same way, promoting sales of VHP products and ensuring that profits are retained for future outreach activities.

Similarly, USGS policy dictates that flyers and other inexpensive VHP outreach products be distributed free of charge to the public. Some of these materials are in great demand and are given out in large numbers. Paradoxically, the success of these items poses a problem for the VHP because the program bears all costs for their production and distribution. The result is a disincentive to disseminate these popular items because the costs involved constrain other outreach activities of the VHP.

There is a widespread belief within the VHP that staff members who devote significant time to outreach are penalized in the USGS performance review process. Although many public statements extol the importance of outreach, when it comes time for promotions, work in this area is thought to interfere with, rather than help, career advancement of scientists. The committee was told that at present, only senior scientists who have reached the top rungs of the career ladder and expect no further promotions can safely involve themselves with outreach. Those at lower organizational levels hesitate to divert their careers to such nonscientific pursuits. Administrative supervisors have to change this perception by suitably recognizing effective public service. The committee believes that there should be adequate rewards, at all levels, for involvement in outreach activities.

SUMMARY

Crisis response procedures at VHP observatories are well integrated and applied to hazards mitigation. VDAP, while evoking strong praise from the committee, needs to be strengthened, in both personnel and budget. The committee urges wider involvement of VHP personnel in VDAP activities, which—besides providing depth to the VDAP—would permit a wider circle of scientists to gain firsthand experience with volcanoes in crisis. Data gathered during international volcano crises must be better archived and, where appropriate, published. The committee realizes that data acquisition and use can be a sensitive issue with foreign governments and organizations but urges that protocols be explored to improve the ways in which data from one overseas crisis might be better integrated and applied to the next crisis. Existing outreach products of the VHP were judged by the committee to be of high quality and effective in helping mitigate volcano hazards. This effectiveness can be increased by developing ways for the VHP to retain proceeds from the sale of these products and by removing the impediments that limit the involvement of midcareer VHP personnel in their preparation and dissemination.

5

Programmatic and Institutional Issues

Not all of the problematic issues identified during this review could be neatly categorized as affecting only research, assessment, monitoring, crisis response, or outreach. Several of the barriers to effective hazards mitigation influence two or more of these programmatic areas. This chapter considers three such themes that impact the full spectrum of VHP activities: human resources, integration and communication, and priority setting and accountability. Accomplishing the specific recommendations in earlier chapters will not be possible without simultaneously addressing three cross-cutting obstacles: inadequate staff who are inappropriately distributed, lack of coordination between the activities of the VHP and those of the rest of the volcanological and broader natural hazards establishment, and insufficient oversight of what individual scientists do.

HUMAN RESOURCES

Prior to the 1980 Mount St. Helens' eruption, the VHP consisted of the HVO and a program of hazards assessments of the Cascade Range volcanoes; staff for these activities were headquartered in Hawaii and Denver, Colorado, respectively. After this eruption and the funding increase that followed, the VHP greatly increased its presence in the Cascade Range, quickly adding staff and establishing the Cascade Volcano Observatory.

Although both the GD and the WRD added new scientific staff to the VHP, they did so in different ways. The GD transferred a number of existing volcano specialists into the VHP from other USGS programs. This had the immediate benefit of quickly putting in place a team of experienced scientists, but it resulted in the hiring of comparatively few new scientists to the program. The WRD, not having a large cadre of volcano specialists from which to draw, hired a number of young, recently graduated scientists. Many of these WRD hires had been trained in quantitative approaches to the earth sciences, and brought these important capabilities to the VHP.

Currently, the VHP has a large number of capable scientists. However, *the almost total failure of the program to hire more than a token number of new personnel over the past 15 years has created a crisis of continuity in which much of the VHP's accumulated knowledge is in danger of being lost because of upcoming retirements.* These losses, if not offset by future hires, will have serious consequences during future volcano emergencies. Because eruptions are so idiosyncratic and variable, the most valuable asset is firsthand experience with previous events. Overlap of new staff with existing staff is essential for orderly transition of duties and transfer of knowledge, not only of volcanology and associated hydrology, but also of procedures for communicating with users of information. This is especially critical in the case of VDAP, where knowledge of effective crisis management resides in the experience of a small group of scientists and technicians, many of whom may retire in the next 5-10 years. Crisis response also requires energetic people who can work long hours and intensively for long periods— sometime weeks to months. The situation is serious now but will be eviscerating in fewer than 10 years. With the loss of personnel, and no replacements, the domestic response capability is likely to collapse and programs such as VDAP could disappear. **The committee believes that if the VHP does not begin to hire new staff immediately, the program will not be able to maintain response readiness.** The committee suggests that the VHP begin planning for rejuvenation of its work force. This exercise should build upon the program's strategic plan and take into account the new areas of expertise that will be needed in the future.

In many ways, the importance of technicians to the VHP equals that of scientists. Successful mitigation of the effects of the eruptions of Kilauea and Mauna Loa by the HVO in the 1960s and 1970s owed as much to the expertise and creativity of the technical support team as to the accumulated knowledge of the scientific staff. When scientists at the more recently created CVO and AVO were confronted by the challenges of monitoring and responding to erupting volcanoes, it was the experience of the engineers and other technical personnel that helped them cope with the steady stream of emergencies. These individuals have highly diverse backgrounds and in many cases have participated in several decades' worth of crisis response, especially as VDAP has expanded. *The lack of hiring in this area seriously threatens the well-being of the program.*

The committee was told of differences in scientific staffing at HVO, AVO, and CVO. Scientists are permanently headquartered at the latter two observatories, much as they are at any other USGS center. At HVO, however, a significant fraction of the research staff rotates through the observatory on three- to five-year cycles. These people, who historically transferred from other USGS programs, bring new ideas and enthusiasm to HVO and contribute to the VHP's goals. In the past 15-18 years however, most of the scientists who came from other USGS programs remained in the VHP when they rotated back to the mainland rather than returning to their programs of origin. These people thus become permanent additions to the VHP, requiring the long-term dedication of personnel slots and salary dollars that might otherwise be used to hire junior scientists. Besides bringing in fresh and modern perspectives, junior scientists are potentially more mobile. The committee urges the USGS management to acknowledge that the VHP has high priority, so that at least some of the rotating scientist positions at HVO could be filled with new hires. At the end of their HVO tour of duty, these people could be transferred to other parts of the VHP.

INTEGRATION AND COMMUNICATION

In addition to the reduction in employment in the VHP, the past 20 years have seen a change in the relative importance of volcanologists from universities and from other federal agencies. Prior to 1980, the majority of research-active volcanologists within the United States worked for the VHP. Scattered groups and individuals could be found at a handful of colleges and universities, as well as at the Smithsonian Institution in Washington, D.C., and in the national laboratories run by the DOE. Widespread publicity about the Mount St. Helens eruption along with the expansion of global communications made the public much more aware of eruptions worldwide and led to increased student enrollments and faculty hiring in volcanology at universities in the United States and abroad. This trend contributed to a dramatic relative increase of volcanologic knowledge outside the USGS.

One of the implications of this changing balance is that VHP members can no longer rely solely on interactions with other program members to keep them aware of the latest developments in volcanology. Attendance at national and international conferences, participation in professional organizations, and service on editorial boards and review panels have become increasingly important means to stay abreast of how volcano science is evolving. Members of the VHP have highly variable records of involvement in these "extracurricular" activities, partly because of budget constraints and partly through apparent lack of motivation or managerial encouragement. The result is that some VHP scientists appear to have a relatively parochial or obsolete view of their field, making it more difficult for them to carry out their responsibilities effectively.

Even if the number of VHP employees were to increase over the next few years, it would probably be insufficient to keep up with new techniques and with the increased flow of scientific knowledge that threatens to overwhelm the already overworked VHP staff. The resulting shortage means that the VHP will have to either reduce the scope of its mission (in conjunction with major retraining of existing staff) or increase the pool of workers who can help it accomplish program goals. Because of this situation, *the committee concludes that the VHP can no longer accomplish all of its goals through in-house activities.* **The committee recommends that to accomplish its goals, the VHP increase its coordination and collaboration with researchers from other parts of the USGS, other federal agencies, academic institutions, and industry.** Although many VHP scientists today have good collaborations with scientists outside the USGS, others appear to avoid such interactions. It is in the interest of VHP management to more strongly encourage this sort of endeavor.

The VHP could take several steps to better accomplish its goals through enhanced academic collaboration. The VHP could colocate more staff and facilities at universities. Elsewhere within the USGS, colocation of offices on university campuses or mere proximity to academic centers offers excellent rewards for both sides. Examples of successful colocations are found at the University of Arizona, California State University at Sacramento, and the University of Washington. The Menlo Park office of the USGS has maintained close working relationships with students and faculty at Stanford and at the University of California, Berkeley, for more than four decades. Among the three AVO partners, most difficulties appear to be caused by the large distance between the USGS office in Anchorage and the University of Alaska and State of Alaska offices in Fairbanks. The committee heard surprisingly little about cooperation between HVO and the strong volcanological program at the University of Hawaii in Honolulu. Possibilities for a stronger link also exist between CVO and the University of Washington, which is currently responsible for seismic monitoring throughout the region. The committee saw a missed opportunity in the apparent failure of a recent proposal to colocate CVO with a new campus of Washington State University in Vancouver.

A final communications issue surrounds the way the VHP interacts with civil defense officials. The committee heard a few comments about this aspect of the program. Senior VHP administrators explained that they felt strongly that their role was to provide scientific background necessary to help public officials make policy judgements associated with volcanic hazards, but not to get directly involved in the decision-making process itself. The committee understood and mostly concurred with the reasoning behind this separation of tasks between the VHP and the local government agencies. On the other hand, the VHP should be more aggressive in promoting the economic benefits associated with its mitigation activities, both to upper management within the Department of the Interior, and to Congress.

Students

The HVO, CVO, and AVO have served as informal training facilities for small numbers of students throughout the past several decades. Most of these students either have been volunteers or have come with support from their home institutions. The VHP could derive many benefits from involving more students at all levels in the daily operations of its observatories. Students can join field crews, compile and archive data, and help with public education and outreach, all while working on their own research projects. For the student, it is a unique opportunity to get hands-on experience in volcanology. HVO has long assisted small numbers of students from the mainland and from other countries to work on Kilauea and Mauna Loa. Many Latin American volcano observatories also make effective use of students. The British Geological Survey was recently successful in getting British graduate students involved in monitoring the Soufriere Hills eruption on the island of Montserrat. Because many of these students are likely to be future leaders in volcanology, such programs should be expanded. Undergraduate and graduate students can participate through cooperative programs, temporary-contract hires, volunteer positions, and so forth.

Postdoctoral scientists, particularly those with primary training in engineering, computer science, chemistry, or some other discipline besides volcanology can offer new ways of looking at monitoring problems. Others with backgrounds in the social sciences and public policy could help VHP scientists craft more effective crisis response protocols. A vigorous postdoctoral program not only would bring bright young scientists into the VHP, but would also forge stronger links with the university community.

Extramural Grants Program

Traditionally, the USGS in general and VHP in particular have differed from other federal science agencies in having almost all of their research conducted in-house by members of the organization. This arrangement worked well when the USGS budget and staff were growing. As discussed elsewhere in this report, such is no longer the case. For example, monitoring activities at observatories can take so much staff time that there is little left for the scientific investigations that may lead to new monitoring approaches. An extramural grants program (perhaps modeled after the interagency National Earthquake Hazard Reduction Program) would take advantage of the talents of outside researchers in academic, government, and private institutions and focus their efforts on VHP goals. By influencing the types of projects that are approved, the VHP could direct university researchers toward those problems that would best complement ongoing program activities. The committee recognizes the challenges of implementing such a program under current budget constraints and in the absence of the infrastructure to administer it. However, an investment in this area would allow considerably more research to be carried out on problems that are most relevant to program goals.

Personnel Exchanges

A sabbatical or Intergovernmental Personnel Act (IPA) program that assigned VHP personnel to universities or other federal agencies, such as NASA, NOAA, the Smithsonian Institution, and Los Alamos National Laboratory, for periods of several months to a year would be another means to increase collaborations with outside entities. In exchange, university faculty on sabbatical leave could bring the latest concepts to observatories, serve on VDAP deployments abroad, or participate in outreach activities.

Federal Coordination

A broader communication problem exists among all of the federal agencies involved with volcano hazards research. Communication among these groups seems to be more ad hoc than systematic, with little apparent coordination from year to year as federal budget requests are prepared. This "balkanization" of U.S. volcanology results in inefficiencies and duplication of effort in the federal establishment. The VHP is urged to be sensitive to this situation and take steps to increase interagency communication whenever possible. This issue could be partially addressed by having regular meetings of volcano-related policy makers within the Washington, D.C., area, including the VHP, NASA, NSF, NOAA, FAA, FEMA, the Smithsonian Institution, and relevant offices of the Departments of the Interior, Energy, State, and Defense.

The committee concludes that there is insufficient integration and communication between the VHP and other government entities involved in volcano hazards. The VHP should take steps to ensure USGS management realizes that the overall scientific goals of the program would be enhanced by such interactions. **The committee recommends that the VHP improve outside communication and better integrate its programs with those of other relevant organizations and government agencies**.

PRIORITY SETTING AND ACCOUNTABILITY

Overall VHP Priorities

The VHP's Five-Year Science Plan for 1999 to 2003 outlines a wide array of activities, ranging from volcano monitoring and crisis response to scientific outreach and information dissemination. If the VHP continues to be faced with flat budgets and limited staff growth, it must prioritize more clearly among these activities. *The committee urges the VHP to put in place a more formal mechanism for prioritizing its activities and seeing that they are consistent with stated program goals.* A possible guiding principle would be to preferentially fund those projects that the VHP is uniquely positioned to carry out, such as volcano monitoring and long-term field studies, while leaving other functions such as small topical research projects and educational outreach to other groups inside and outside the USGS.

The committee questions whether priorities have been set for study of individual volcanoes, or groups of volcanoes, within either the Cascade Range or the Aleutians. The committee gained the impression that in many cases, individual VHP scientists or small groups of scientists select the volcanoes they want to work on and that these projects may continue indefinitely. The committee is not aware that deadlines have been set for completion of these studies, or that the overall approach is more coordinated than haphazard.

A major issue that underlies any discussion of VHP priority setting and accountability is the lack of a clear and consistent management structure. Depending on his or her location and their inclination, an individual VHP scientist or technician might report to one of the four observatory scientists in charge, to the head of the Western Region in Menlo Park, to the local branch chief in Flagstaff, to the VHP coordinator in Reston, or to one of various administrators within the Water Resources Division. This confusing arrangement results in part from the Geologic Division's tradition of using rotational administrative assignments, rather than hiring or developing career administrators. Frontline VHP scientists have long been expected to serve in a managerial position for a few years and then return to continue their research. An advantage of this approach is that the people in charge retain firsthand familiarity with the issues affecting their staff. However, other scientific organizations (including the WRD) have long recognized that effective research administration is commonly incompatible with the maintenance of a vigorous individual research portfolio and thus encourage a longer-term commitment to a nonresearch career. The main drawback of the current complex structure is that it creates an institutional barrier to the emergence of strong leaders. This lack, in turn, makes individual staff members unsure about who sets their priorities and makes the VHP as a whole less influential within the prioritization and budget-setting processes of the USGS and DOI.

Observatory Priorities

Because most staff members of the VHP report to one of the scientists in charge of the four volcano observatories, these four individuals have special responsibilities for setting, assessing, enforcing, and coordinating prioritization across the program. In the observatory environment, volcano monitoring, hazard assessment, and communication with civil authorities may be most important, but during periods of volcano unrest and newly evolving activity, volcano crisis response assumes special priority. For example, the 1983-2000 (and continuing) eruption of Kilauea volcano in Hawaii must be monitored, but the committee suggests that the levels of funding and personnel invested in this noncrisis event should be balanced so as to not preclude other observatory studies of this still incompletely understood volcano.

An important aspect of priority setting relates to the timeliness of scientific publication. As VHP priorities evolve, managers can be tempted to move a given scientist from one project to another before he or she has written up for publication the results from a previous research assignment. Scientific publication is an important end product of VHP research, not only for the needs of civil authorities but also for other scientists (both USGS and non-USGS) who benefit from additions to the literature on volcanoes and volcano products. The problem is particularly acute when unpublished studies involve volcano hazard assessments that could have a direct bearing on the safety of people and property. *The committee urges that higher priority be given to the timeliness of scientific publication.*

As mentioned earlier, many VHP staff members do not appear to be very involved in professional associations, beyond attending meetings of the American Geophysical Union and the Geological Society of

America. Such participation can be particularly important in developing collaborations with volcanologists outside the VHP and in highlighting the importance of the VHP to the larger volcanology community. This is another area in which the committee believes that prioritization is important. Scientists in charge must give more clear direction about the balance of professional activities of VHP staff members. Scientists in charge and the rest of USGS management should encourage and reward greater involvement of VHP members in the non-USGS professional scientific community.

The committee recommends that the VHP set priorities and review them periodically at all levels from program to observatories to individual performance plans. For example, the VHP should produce a prioritized list, with completion dates, for all volcanoes of the Cascade Range for which comprehensive research projects and hazard assessments will be conducted. The ongoing collaborative research program at Mount Rainier volcano serves as a good example. Rather than wait until comprehensive research projects are completed at individual volcanoes and the supporting data and research results are published in the refereed literature, preliminary or less comprehensive hazard assessments should be prepared for as many domestic volcanoes as possible, with the expectation of future revisions and publication of supporting documentation. Such assessments have recently been completed for the 10 large volcanic centers in Washington and Oregon and for several in the Aleutian Islands.

Interdivisional Issues

From the late 1960s until Mount St. Helens erupted in May 1980, the GD administrated the VHP and carried out all programmatic investigations. Soon after the Mount St. Helens event, VHP managers realized that scientists of the WRD, with their expertise in volcano-related hydrologic and sediment transport processes, could contribute much to VHP goals. Accordingly, the program funded a number of WRD projects, and the two divisions worked together as a single team for about 18 months.

In the 1980s, disagreements between the two divisions prompted the USGS director to partition the VHP into two components. This division in effect created two programs, staffed and operated separately, based on different floors of the same building. The coordination between these two parts of the program has waxed and waned in effectiveness as managers from the two divisions have changed with time. The percentage split between the divisions, however, has hardly varied, being apparently unrelated to evolving programmatic goals and opportunities (Figure 1.5).

The VHP funding split between the WRD and GD has produced tensions and perceived inequities. It is questionable whether scientific investigations and results throughout the program are integrated as effectively as they could be. The VHP is a USGS program and should be operated in ways that foster seamless relationships among staff within the GD and WRD. **The committee recommends that USGS management integrate the GD and WRD parts of the VHP.** Annual competition for project-level funding should take place within the same arena, free from divisional setasides. Whatever solution is implemented, it must meet the needs of the program, foster both applied and process-oriented research, and appropriately reward employees for published research, assessment and monitoring studies, and public outreach. Such a shift would have to recognize the existing divisional differences in personnel practices and provide a means for orderly transition for those individuals who would change their affiliations.

Another aspect of the lack of integration is that, in some respects, the VHP appears to be a budget line item rather than a program. There is competition between the observatories and the divisions. Both have personnel who report to either GD or WRD. The committee hopes that integration of the GD and WRD components of the VHP will help resolve issues of prioritization and allow for stronger centralized management.

Data Access and Management

The telecommunications explosion in the late 1990s has been accompanied by fundamental shifts in the way scientists view the data they gather and the way institutions such as NSF, the Congress, and the public view the use of such data collected with public funds. In the past, most such information was considered proprietary, and researchers were highly possessive of what they collected until it could be incorporated into publications. Members of some of the most collaborative scientific fields, such as astronomy, particle physics, and planetary geology, were the first to reverse this pattern, posting their data to the Web, initially in a trickle but then in a flood. To make this practice work, they needed to develop protocols for formatting, documenting, storing, and interpreting the data.

Volcanologists, both in academia and in the USGS, have been relatively slow to adopt this approach. As a result, data management by different observatories and different individuals within the VHP continues to be idiosyncratic and inconsistent. The main problem this creates is that it complicates attempts to do either comparative studies of different eruptions at different volcanoes or longitudinal studies of a series of eruptions at a single volcano. A second problem is that users of volcanic information, such as civil defense authorities, the press, and insurance companies, are unable to get a consistent picture of hazards and the processes that cause them. Third, poor data documentation and availability limit studies by non-USGS scientists. A fourth problem is this situation hinders collaboration with volcanologists outside the VHP.

Data documentation, storage, and access are major evolving issues throughout the scientific community. Consensus is emerging that (1) public access to most data collected with public funds is essential; (2) proprietary rights of individual investigators should be respected, but the duration of exclusive rights should be limited; (3) providing documentation and access requires substantial investment of funds and personnel; and (4) neither the scientific community nor research institutions recognize or reward the production of high-quality, well-documented, publicly accessible data sets. Adherence to points 1 and 2 by the VHP and, indeed, much of the scientific community has been limited because of points 3 and 4. The reasons are obvious: lack of funds, staff, rewards, and in some cases, equipment.

Examples of effective data management systems exist, such as IRIS for seismic data and UNAVCO for GPS data. However, these examples have been developed in scientific cultures where consistent data formats are more common than in volcanology. In the case of earthquake data, collection and coordinated management of information from geographically dispersed sites are essential for much of the work of the field, so motivation is high. The VHP and VDAP in particular face much more complex challenges, including working in multidisciplinary and even crisis situations of long duration.

Standardization of data management protocols and formats across observatories and VDAP deployments is essential in order to improve access for the scientific community and others. The committee believes that the potential benefits of public access outweigh the possible drawbacks of data misuse. Also at issue is the need to clarify the rights, responsibilities, and rewards for timely data posting by individual investigators, observatories, and the VHP as a whole. **The committee recommends that the VHP set standards for documentation, archiving, and access policies, including the length of the proprietary period.** Once the data are compatible, the VHP should link the holdings of different observatories, perhaps through a virtual data center on the Web, and make them available to other scientists and the general public.

Although old data sets can be extremely valuable, transforming them to usable form can be a daunting challenge. The committee recommends a structured program of prioritized resource allocation for bringing legacy data sets to high-quality, well-documented status. The personnel performance review and rewards system within the VHP should recognize the importance of high-quality, well-documented, publicly accessible data sets.

The committee has not made specific recommendations as to the timing of data release, but suggests that standards be flexible and consistent with those promoted by the National Science Foundation or those used by other portions of the USGS such as the Earth Hazards Reduction Program. The committee also does not specify the types of data that should be released. However, the committee would like to see as many data sets as possible released, including but not limited to tiltmeter, GPS, seismic, gas, and thermal data. Very recent discussions within the volcanological community about standardized data collection and dissemination, including a workshop scheduled for the IAVCEI meeting in Bali, Indonesia, in July 2000, should receive careful attention from the VHP. One of the side benefits of improved information management, especially timely access to public data, will be to enhance the leadership roles of VHP scientists within the larger volcano research community.

6

A Vision for the USGS Volcano Hazards Program

The committee hopes that if a major eruption were to occur in the United States in the year 2010, the USGS VHP would be prepared to respond in a manner more similar to Prologue 2 than to Prologue 1. This chapter discusses in greater detail the committee's vision for the VHP exemplified in Prologue 2.

Most of the technology and understanding described in this vision exist today, and others are extrapolated from current research. The scenario is optimistic but realistic. For this country to have a VHP capable of saving tens of thousands of lives and greatly limiting property damage and economic disruption, investment in technology, people, and basic research is required. Although the hypothesized eruption could occur tomorrow, in 2010, in 2110, or in any future year, the decisions made today will greatly affect VHP's ability to forecast, monitor, predict, and minimize the effects of devastating volcanic events. The consequences of failing are great, as outlined in Prologue 1. The difference may range from a few fatalities to thousands of deaths, from major damage to structures to complete regional economic collapse.

AN ALTERNATE SCENARIO FOR THE 2010 ERUPTION OF MOUNT RAINIER

The first sign that Mount Rainier was reawakening came from crustal deformation measurements at the CVO. A dense array of permanent GPS receivers in the area picked up subtle ground movements nearly a year before the eruption. The motions were so small (<1 mm) that they would not have been recognized by looking at the data from an individual instrument. Only sophisticated computer programs especially designed to integrate numerous data sets and to search for patterns of ground deformation diagnostic of magma migration were able to detect the early warning signs.

Data from a constellation of orbiting InSAR satellites were also critical. Interferometric maps of ground motion automatically generated from these data, with a horizontal spatial resolution of 10 meters and a vertical resolution of 0.1 mm, also showed subtle indications of inflation of a deep magma body. Because these InSAR systems employed longer-wavelength signals than the radars used at the end of the twentieth century, they were less affected by vegetation growth and other surficial processes. Imaging from multiple satellites allowed three-dimensional vector displacements to be determined and compared directly with the GPS measurements. The analysis software detected inflation of part of a deep magma chamber 15 km beneath the earth's surface.

The deformation alert triggered several immediate actions. After presenting the results to the scientist in charge of CVO, the deformation group began an intensive series of tests to check the validity of the data and the automated computer modeling. The scientist in charge brought up on a computer screen a three-dimensional, interactive hologram, showing version 15.3 of the hazard assessment for Mount Rainier and called in the heads of the other scientific groups. The scientist in charge did not have to be reminded that Mount Rainier posed the highest risk of any volcano in the continental United States. Rainier had been a high-priority volcano for study since the 1990s, and hazard assessments had been revised and updated numerous times since then. The nominal annual probability of an eruption with a Volcanic Explosivity Index (VEI) greater than or equal to 2 was 4 percent. (Such an eruption would be classified as explosive and involve roughly one million cubic meters of ash.) The new information would increase this probability significantly.

All major Cascade volcanoes had updated hazard assessments based on extensive field mapping by teams of experienced geologists. Field mapping, combined with subsurface imaging using ground penetrating radar and shallow seismic methods, allowed the hazard assessment team to develop complete three-dimensional maps of deposits from past eruptions, debris flows, and landslides and to assemble detailed descriptions of the volcano's eruptive style. Advances in geochronology during the previous 10 years allowed the team to date eruptive products with unprecedented precision and to derive sophisticated models of processes active within the magma chambers deep below the mountain. The three-dimensional mapping combined with accurate dating and dramatically improved models of volcanic eruption processes allowed the team to assign probabilistic estimates for impending eruption scenarios. The scientist in charge was thus able to update the likelihood of these scenarios in near real time, as monitoring data from satellites and remote locations on the volcano streamed into the observatory. The scientist in charge scanned three-dimensional displays showing the extent of a worst-case scenario directed blast and evaluated the probabilities.

At this point, none of the other monitoring systems reported anomalous signals. Seismicity was at background level and no unusual gas emissions had been detected. The scientist in charge asked for updated seismic and electromagnetic images of the subsurface. The permanent broadband seismic network was augmented with portable stations and electromagnetic sensors brought to the area by university scientists working with researchers from the USGS VHP. Together these scientists were able to create high-resolution images of subsurface structure and time-dependent changes in rock and fluid properties. Laboratory calibrations were used to interpret the images of seismic velocity, attenuation, and electrical resistivity in terms of temperature, composition, and extent of partial melting. Based on these results, the team reported to the scientist in charge that the deep magma chamber beneath the volcano had indeed swelled in size and changed in shape. Given the available data, they were able to estimate the size, shape, depth, and location of the magma body. The potential for an eruption was identified, but the future behavior was still unclear.

By this time, civil defense officials, as well as state, county, and municipal authorities had been briefed on the changes taking place beneath the mountain. Although there were no visible signs of unusual activity, these officials were well aware of the potential hazards posed by the volcano. VHP scientists had been working closely with Washington State emergency management personnel. At this point the scientist in charge reported that it was too early to determine whether this increase in the size of the magma chamber would lead to an eruption. The USGS issued an updated probabilistic assessment based on the new geophysical data. The Washington State Emergency Management Agency developed an action team to deal with a worst-case Mount Rainier eruption scenario. The action team reviewed specific plans and assignments developed over the past decade in the following areas:

- identification and mapping of the hazard zones; registering of valuable movable property;
- identification of safe refuge zones to which the population could be evacuated;
- identification of evacuation routes, their maintenance, and clearance;
- identification of assembly points for persons awaiting transport for evacuation;
- means of transport and traffic control;
- shelter and accommodation in the refuge zones;
- inventory of personnel and equipment for search and rescue;
- hospital and medical services for treatment of injured persons;
- security in evacuated areas;
- the formulation of alert, warning, and evacuation procedures;
- relocation and recovery activities; and
- provisions for revising and updating the plan.

During this time, a university team brought in an array of absolute gravity meters, capable of measuring the earth's gravitational force with an accuracy of 1 part in 100 million. They detected a slight increase in the gravity field that, when combined with the GPS, InSAR, seismic, and electromagnetic imaging results, helped constrain estimates of the density and composition of the magma. The results were not encouraging. The magma was in all probability dacitic, the same composition as the devastating eruption of Mount St. Helens in 1980. At the same time, field teams began monitoring the volcanic edifice in response to the subtle changes in strain. Special care was taken to monitor local strain rates in well-known alteration zones high on the flanks of the volcano in order to record the response of the shallow hydrothermal system to the new activity.

Meanwhile VHP scientists used geophysical and geochemical observations to initialize eruption models that predict the volume of erupted material, height of ash plumes, size and distribution of pyroclastic flows, and related hazards. Based on recent improvements in understanding magma rheology, chemistry, and volatile kinetics, it was possible to integrate the physical and thermodynamic governing equations forward in time to predict magma flow, eruption potential, and behavior during eruptions. Although these models were not sufficiently well constrained to accurately predict the exact time and magnitude of the eruption, they did reveal a range of outcomes that were then considered by disaster planners. The predictive capability improved as more data were collected and the volcanic activity increased.

Indeed, much of the improvement in the ability to assess and forecast volcanic hazards had come from an improved understanding of volcanic processes. These advances were derived from basic research in theoretical, numerical, and laboratory studies, along with knowledge gained from global monitoring of volcanic systems. Advances in understanding physical and chemical processes improved monitoring capability and led to better methods for interpreting data. At the same time, the integrated data sets collected at active volcanoes and in the course of hazards assessment studies provided the basis for testing concepts about processes within active systems. One practical result of this research was the construction of engineered barriers around Mount Rainier, capable of diverting flows of mud and debris away from critical facilities and population centers. Structures were strengthened to withstand anticipated ash accumulations.

At the same time, other scientists were analyzing the stability of the glacier-clad flanks of the volcano and the possibility of devastating debris flows. Computer models of hot ash accumulation onto Rainier's snow and ice fields estimated the maximum possible runout distances and flow volumes. Acoustic flow monitors, which detect the ground vibrations due to fast-moving mudflows, were double-checked, and people in the path of possible mudflows were alerted and evacuation procedures reviewed. These models and monitoring devices had been developed in part based on sophisticated studies of debris flows using a large-scale experimental facility designed and run by USGS VHP scientists.

Six months after the first signs of magma chamber swelling, the first unusual seismic events were recorded. A dense, broadband, high-dynamic range, seismic network jointly operated by CVO and a consortium of university groups detected an intense swarm of earthquakes at a depth of 8 km. This was considerably shallower than magma sources previously noted and was the first indication that magma was migrating upward. Within a few weeks, the first long-period seismic events were recorded, providing further evidence of magma flow. High-resolution earthquake locations using full seismic waveforms yielded resolution of a few meters in near realtime, allowing for precise imaging of seismically active structures. The earthquakes outlined three fingers of magma migrating upward through the crust. Sophisticated source modeling helped seismologists locate constrictions in the magma conduit that caused pressure increases followed by episodic discharges of magma. Shortly thereafter, the first recordings of harmonic tremor, a low-frequency oscillation detected by seismic instruments, were reported. Research by VHP postdoctoral scientists working with VHP theoreticians had led to well-tested models of harmonic tremor. Arrays of seismic instruments were able to locate the source of the tremor, and the amplitude information was used to calibrate the size of the conduit and the volume flux of melt.

The VDAP team installed a network of five borehole tilt and strainmeters to 100-meter depths using microdrilling methods. The VDAP team had decades of experience in volcanic crises throughout the world, and most VHP personnel had invaluable hands-on participation in VDAP. The miniaturized instruments, using sensor-on-a-chip technology developed in partnership with the Department of Energy, began monitoring strain signals associated with the pulse-like rise of magma within the conduit system. The VDAP team was prepared to handle not only the voluminous influx of monitoring data, but also the increasingly aggressive media attention that the volcanic awakening had created.

By this time, the USGS had issued its first low-level eruption forecast. The time and the magnitude were still unclear, but the potential for an eruption grew increasingly more likely. The projections were updated periodically as more data became available, in much the way that weather forecasts were done at the end of the twentieth century. Seismic, strain, and geochemical data were available in near real time over the Internet. Because of extensive outreach and education, the public was generally able to manage the information flow and make reasonable decisions. Many residents living within the highest hazard zones decided to leave the area. Disaster preparedness planning was well advanced. The Washington State Emergency Management Agency task force completed a tabletop simulation of the Mount Rainier eruption emergency plan. Modifications to the response plan were made based on lessons learned from the exercise.

Eleven months after the first indications of ground deformation and five months after the onset of increased seismic activity, the first signs of volcanic gases were recorded. Miniaturized gas sensors placed on broad fracture systems on the sides of the volcano picked up emissions of CO_2 and helium. In addition to noting high concentrations of these gases, the in situ sensors were able to measure their isotopic compositions, which showed a clear magmatic signature. At the same time, space-based sensors, developed in partnership with NASA, including LIDAR, with the capability of monitoring gas compositions, revealed very low amounts of gas being released from the volcano. This was particularly alarming, because the estimates of magma volume, from geodetic, strain, gravity, and seismic tomography indicated a large migrating body. The absence of gas emissions suggested that the magma was not degassing at the surface, and therefore was building in pressure.

By this time the governor of the State of Washington had ordered the evacuation of the volcanic hazard zones. This was based on information from the VHP personnel estimating a high probability of an explosive eruption of Mount Rainier during the months of June or July. Although air traffic was routine, an official notice to aircraft flight personnel from the FAA and the Volcanic Ash Advisory Center was released that described the potential for an eruption to inject large amounts of ash into the atmosphere. Existing contingency plans for rerouting air traffic were implemented. Shortly thereafter, the president of the United States declared a state of emergency for Washington State. This made federal resources available to aid the state and its affected residents. The governor and president acknowledged the tremendous coordination of scientific information by the VHP, not only to public safety officials, but also to the affected population. This coordination minimized confusion about what to believe and whom to believe. On May 15, 2010, the governor of the State of Washington announced the completion of the evacuation of hazard zones identified by VHP personnel.

On May 18 the eruption process accelerated exponentially. Until this time the effects had been subtle. Indeed, none of the many signs tracked by USGS scientists would have been detected without sensitive instrumentation, developed over the last 10 years in collaboration with numerous university and government colleagues. Beginning at 1:00 p.m., an intense swarm of earthquakes started, with hypocenters shallowing markedly over the next two hours. Harmonic tremor amplitude increased dramatically. Automated event location algorithms identified zones of intense fracturing ahead of a rising magma body. Strain-measuring instruments and GPS receivers recorded motions of meters in a matter of hours as the dike rose toward the surface. At this point, a short-term forecast of high probability of an explosive eruption was issued. Air traffic was diverted away from the region and critical facilities went into automated shutdown procedures. The Air National Guard was called in to deploy a widely dispersed network of relatively inexpensive, case-hardened, biodegradable microsensors capable of detecting and relaying, via satellite, ambient temperature, pressure, humidity, and geochemical conditions to recording stations tens of kilometers away. This technology, developed for use on battlefields, had recently been partially declassified for application to natural hazard emergencies.

The sequence was unusual in the rapid acceleration toward the climactic eruption. At 3:00 p.m. the north flank of the volcano exploded in a directed blast. This was followed shortly by an eruption cloud rising to 30 km in the atmosphere. Seismic and acoustic infrasound networks combined with space-based optical, radar, and thermal sensors, operated by NOAA, NWS, NASA, and nuclear treaty monitoring networks, rapidly detected the onset of the eruption, thereby broadcasting an instantaneous notification around the world. A combination of these data and the ambient information provided by the widely dispersed microsensors was used to rapidly quantify the size and explosivity of the eruption. Numerical models of ash dispersal combined with accurate models of wind direction and strength, operated in collaboration with the NWS and the FAA, provided timely warning of ash hazard to aircraft and other affected entities.

As it rose, part of the volcanic ash column became unstable and collapsed, producing pyroclastic flows that raced down the Puyallup and Carbon River valleys. Within hours, debris flows inundated other areas low on the flank of the volcano. Some of these debris flows were diverted by engineered structures, built low on the flanks of the volcano for this purpose. Property damage and loss of life in these areas were limited. Elsewhere, engineered systems failed to contain the debris flows. Residential areas and business parks in the lower Nisqually River were devastated when the Alder Dam failed in response to a lahar pulse entering the reservoir around 5:00 p.m. Fortunately, these events were anticipated by hazard assessments and, as a result, by the public. Even in these areas, loss of life was minimal because of prompt evacuation by communities well aware of the risks. Property damage, however, exceeded several billion dollars.

As the eruption progressed through the night, VHP staff focused on forecasting rates of ash accumulation in nearby communities, continued lahar hazards, and the probable duration of the eruption. All of this information was crucial in the following days to organize a safe and rapid response to the disaster.

7

Principal Conclusions and Recommendations

This chapter summarizes the principal conclusions and recommendations developed elsewhere in this report. Major conclusions are printed in italics and recommendations in bold.

The VHP is comprised of a dedicated scientific and technical staff that has a wealth of practical experience, coupled with good theoretical understanding of underlying volcanic and hydrologic processes. To help society prepare for and deal with the effects of volcanic eruptions, the VHP uses five interrelated approaches: (1) long-term *hazard assessment,* (2) *monitoring* baseline measurements that allow premonitory changes to be recognized, (3) *crisis response* when a volcano is erupting, (4) topical studies of geologic processes that allow for better understanding of the causes and consequences of volcanic hazards, and (5) communicating with civil authorities and the surrounding communities about the results of their studies. These five approaches all aim to help society respond to the dangers posed by volcanoes. Another way to view these activities is to consider a continuum of three overlapping types of societal response to eruptions: research (knowledge acquisition), operations (knowledge application), and outreach (knowledge translation). Research provides the basic information and concepts that underlie the various methods of volcano data collection and interpretation.

The committee was asked to address two questions: (1) Do the activities, priorities, and expertise of the VHP meet appropriate scientific goals? (2) Are the scientific investigations and research results throughout the program effectively integrated and applied to achieve hazard mitigation? The committee's views with respect to these questions are summarized below and at the end of Chapters 2, 3, and 4.

Basic research in the VHP, although reasonably well integrated, is being threatened by budgetary and personnel constraints, which may diminish the program's ability to meet appropriate scientific goals. If these problems are not solved, the program will likely be forced to reduce levels of in-house basic research and/or to increase collaboration with non-USGS scientists. *Hazard assessment,* while traditionally strong in geologic mapping, radiometric age dating, and related activities, has to be strengthened in modeling and probabilistic approaches if the program is to continue to meet appropriate scientific goals. Existing hazard assessment activities at individual volcano observatories are effectively integrated and applied to hazard mitigation issues. The one-volcano, one-scientist projects under way at some volcanoes, although scientifically appropriate, may not be effectively integrated with each other or with the VHP as a whole.

Continuing budgetary pressures place four types of constraints on the VHP's ability to *monitor* volcanoes. (1) Aging equipment is not replaced soon enough (or at all), increasing the chances of failure during a crisis. (2) The VHP's traditional role as the developer and tester of new monitoring equipment and techniques is jeopardized. (3) The number and extent of regular instrumented surveys, which are crucial for the success of any monitoring program, are restricted. (4) Personnel familiar with new techniques are not hired. If the current situation is not reversed, the VHP may not be able to field the best instruments or to maintain its traditional high standards for monitoring. These issues apply to varying degrees to all of the monitoring methods used by the VHP, and if they are not addressed in the near future, the program runs the risk of not being able to meet appropriate scientific goals. On the other hand, the monitoring methods currently employed in the VHP seem to be well integrated and applied to achieve hazards mitigation.

Crisis response procedures at VHP observatories are well integrated and applied to hazards mitigation. The VDAP, while evoking strong praise from the committee, has to be strengthened, in both personnel and budget. The committee urges wider involvement of VHP personnel in VDAP activities, which—besides providing depth to the VDAP—would permit a wider circle of scientists to gain firsthand experience with volcanoes in crisis. Data gathered during international volcano crises must be better archived and, where appropriate, published. The committee realizes that data acquisition and use can be a sensitive issue with foreign governments and organizations but urges that protocols be explored to improve the ways in which data from one overseas crisis might be better integrated and applied to the next crisis. Existing outreach products of the VHP were judged by the committee to be of high quality and effective in mitigating volcano hazards. This effectiveness can be increased by developing ways for the VHP to retain proceeds from the sale of its products and by removing impediments that limit the involvement of midcareer VHP personnel in their preparation and dissemination.

RESEARCH

It is difficult to separate the contributions to basic volcanological knowledge made by VHP scientists from those made by their colleagues in other parts of the USGS, other government agencies, universities, other countries, and the private sector. Nonetheless, throughout much of the second half of the twentieth Century, members of the present-day USGS Volcano Hazards Program were national if not global leaders in the formulation of ideas about how volcanoes work.

The committee did not review individual VHP research projects, nor did it conduct an in-depth assessment of the research component of the program. However, the committee feels strongly that USGS management must ensure that most, if not all, basic research projects are directed toward program goals. Such assurance can come from internal USGS programmatic oversight and from careful structuring and enforcement of the annual performance plans of individual research scientists.

Basic research in the VHP is being threatened by budgetary and personnel constraints, which may diminish the program's ability to meet appropriate scientific goals. One of the most important long-range issues that the VHP must face is deciding how central in-house basic research will be to its mission in the future. Such research is also being done at universities, government labs, and non-U.S. institutions. Thus, one might argue that the VHP could forgo its basic research activities without this having a major impact on the state of knowledge of volcanic processes. On the other hand, eliminating this program element altogether would likely damage the intellectual vitality of the VHP and make it more difficult (if not impossible) for the program to hire topflight young scientists. *The committee believes that if the VHP is faced with continuing budget shortfalls, it could elect to reduce fundamental research activities and redirect scarce resources to monitoring and crisis response functions, which it is uniquely positioned to do* (see Chapters 2 and 3). *However, these savings would come at a high cost.* The ability of the VHP to respond to volcanic crises would be compromised by a lack of expertise in hazard assessment or volcano process studies.

One possible solution would be for VHP members to collaborate more on research projects with scientists outside the USGS, particularly those from universities and from laboratories of other government agencies. More active collaborations, coupled with an extramural grant program for academic researchers overseen partly or completely by the VHP, would help ensure that more investigations that are directly relevant to the program's mission would be carried out.

HAZARD ASSESSMENT

Volcano hazard assessment aims to determine where and when future volcano hazards will occur and their potential severity. This kind of appraisal provides a long-term view of the locations and probabilities of large-scale eruptions and related phenomena, such as volcanic debris avalanches and tsunamis. The extensive range of hazards that must be evaluated requires the combined knowledge of a broad array of scientists, including geologists, hydrologists, geotechnical engineers, atmospheric physicists, and statisticians. *Because assessment is inherently interdisciplinary, the VHP needs access to a very diverse set of expertise, either within its own ranks or through collaborations with outside groups.*

Geologic mapping, stratigraphy, geochronology, and physical volcanology provide the backbone of volcanic hazard assessments by revealing past trends in eruption timing, volume, and explosivity. Historically the USGS has done an excellent job of incorporating geologic data into its assessments. *The committee commends VHP efforts to integrate findings of geologic studies into volcanic hazard assessments.* An ongoing challenge is to more effectively quantify geologic data in ways that optimize their use in such assessments.

Although mapping and dating of volcanic deposits can provide a good framework for hazard assessment, mechanical models of physical, chemical, and hydrologic processes help refine forecasts of the types and magnitudes of future eruptions. Both numerical models and laboratory simulations can relate the boundary conditions on a volcano to the likely consequences of any incipient eruptive activity. Although there has been some VHP participation in the development of these models, especially those related to hydrologic and sedimentologic phenomena, most have been created by non-USGS scientists. *The committee encourages the VHP to include more theoretical modeling of volcanic phenomena in its hazard assessments.*

Because it is impossible to predict eruptive behavior with certainty, particularly for dormant volcanoes, most hazard assessments are inherently probabilistic in nature. Use of three approaches to hazard assessment—mapping and dating, theoretical modeling, and probability calculations—by the VHP reflects the training of its participants. Probabilistic approaches are relatively recent additions to the VHP assessment repertoire, but they have received more attention lately because of their obvious utility in communicating with civil defense authorities and the general public. *The committee strongly encourages the VHP to develop a balanced assessment program that takes advantage of the full range of techniques available to volcanologists today.*

Assessment priorities vary from observatory to observatory, reflecting local differences in the nature of the volcanic hazard and the expertise of the resident scientists and technicians. Volcanic ash interaction with jet aircraft poses the greatest danger from Alaskan volcanoes, because ingestion of ash can result in engine damage or failure. *Although responsibilities for monitoring and crisis response in Alaska are shared among the VHP, the NWS, and the FAA, only AVO is capable of (1) establishing the historical context of future explosive eruptive activity, (2) providing advance warning of an impending eruption, and (3) conducting ground monitoring that can confirm an eruption is actually in progress.* Because of the nature of these dangers, AVO has placed greater emphasis on monitoring and crisis response than on long-term hazard assessment. Only a few of the Alaskan volcanoes have even rudimentary hazard maps. The expense and logistical difficulties associated with access in Alaska preclude the kind of comprehensive mapping strategy carried out by CVO and HVO. Recent AVO-coordinated mapping campaigns at selected Alaskan volcanoes carried out by teams of USGS, other government, and university geoscientists have expanded the coverage of hazard assessment products. *The committee concludes that basic yet rapid assessment of the eruptive histories of as many of the Aleutian volcanoes as possible is necessary to guide prioritization of the placement of instruments used to provide warnings to pilots and other nearby infrastructure.*

If faced with a continued flat budget, the VHP must find ways to carry out its mission more efficiently. **The committee recommends that** the VHP initiate a form of collaborative prioritization with respect to hazard assessment. This might include a broader application of the team approach now being used at AVO and CVO. In addition to prioritization, *volcano hazard assessment within the VHP would be improved by greater consistency of data collection, storage, presentation, and inter-pretation.*

MONITORING

To be effective, monitoring must be done before, during, and after eruptions and must be integrated with carefully designed communication schemes. It requires the type of long-term commitment of time and resources that academic and industry scientists generally cannot make. Furthermore, the quality of monitoring depends on the amount of experience of the participating scientists. For these reasons, the VHP is uniquely qualified within the United States to carry out volcano monitoring.

The combined seismic-deformation approach, which has traditionally been the core of VHP monitoring, tracks phenomena to provide ample warnings of impending eruptions on most volcanoes. The report *Priorities for the Volcano Hazards Program 1999-2003* (USGS, 1999) argues for an expansion of some existing networks and upgrading of overall instrument capability. *The committee endorses these plans because they are directly applicable to the scientific goals of the VHP and will help to achieve hazard mitigation.*

Although there are pros and cons for making data available on a real-time or near real-time basis, the committee believes that the advantages of public access outweigh the disadvantages. The committee therefore recommends that VHP observatories take measures to make their data available on a near real-time basis.

The committee was favorably impressed by AVO's attempts to install seismic networks (either large or small) on as many Aleutian volcanoes as possible. *The committee believes that a team approach for monitoring and studying Aleutian volcanoes from various perspectives should be expanded in the near future so that AVO can provide airlines and other constituents with adequate advance warning of impending eruptions.*

The collection of volcanic gas data is another essential monitoring tool that complements seismic and geodetic information. The committee was disturbed to learn of the paucity of gas geochemical expertise and utilization within the VHP. *The program should reestablish in-house capacity to use and develop both conventional and novel methods for measuring and interpreting volcanic gases.* New ground-based instruments for remote sensing of CO_2 and other gases are currently being developed outside the USGS. These instruments have major technical advantages over existing approaches used by the VHP. *The committee believes that VHP scientists should be in the forefront of such efforts, either by obtaining this equipment themselves or by actively collaborating with groups who are developing these tools.*

Although less prominent in the public's awareness than lava flows or pyroclastic phenomena, mixtures of volcanic debris and water are among the most deadly products of volcanoes. Detection of volcanic debris flows (lahars) close to their sources can provide timely warnings to people in downstream areas. Over the next five years, the VHP plans to improve and field-test remote eruption detection stations for possible deployment in the western Aleutians and the Cascades. *The committee supports this goal because it is relevant to the VHP mission to mitigate volcano hazards.* The VHP should also explore ways to better monitor groundwater flow and pore pressures within volcanic edifices. This type of information could help establish the potential for phreatic and phreatomagmatic activity, sector collapse, and internal pressure buildups capable of generating explosive blasts. *Such hydrologic monitoring warrants greater attention by the VHP.* The incorporation of glacier budget studies as part of VHP monitoring on ice-clad volcanoes would also contribute to this goal.

Another VHP goal that the committee fully supports is the continued development of near real-time remote sensing of volcanoes and their associated ash clouds in areas that are difficult to access. Most of the VHP's remote sensing work is centered at AVO, where satellite data are used to identify thermal anomalies and track eruption plumes and where inclement weather makes traditional observations of volcanoes more difficult. Remote sensing data are becoming integrated only slowly into the monitoring strategies of the other VHP observatories.

A new generation of EOS instruments is now providing potentially useful information for volcano monitoring (e.g., data on thermal regimes, SO_2 gas emissions, deformation, and digital topography). *The committee believes strongly that the VHP should take advantage of this opportunity to the fullest extent possible.* In addition, **the committee urges the USGS to work with NASA to argue in support of an InSAR satellite specifically designed for natural hazards monitoring.**

The committee also considered the potential value to volcano monitoring of two existing remote sensing programs based outside the VHP, the Hazard Support System and the Center for Integration of Natural Disaster Information. The classified nature of the data and the fact that military priorities control which observations are made mean that VHP personnel have limited access and must work through the DOD. This adds an extra bureaucratic layer of communication and interpretation, slowing responsiveness and potentially reducing the effectiveness of the monitoring effort. Second, these programs are very expensive. Thus, the CINDI and HSS initiatives run the risk of draining sparse resources away from the VHP for questionable returns. For these reasons, *the committee cautions against greater involvement with CINDI and HSS unless and until better assurances can be obtained about data access and cost containment. A potentially less problematic alternative would be to establish closer ties with the nonclassified EOS program run by NASA.*

Many volcanoes in the Cascades and several in Alaska lie within wilderness areas and other lands managed by the U.S. National Park Service and the U.S. Forest Service. This situation creates a conflict between the need for effective monitoring in order to serve public interests and the desire to minimize mechanized access to the areas in question. *High-level administrators within the USGS and other organizations must actively campaign to gain recognition that monitoring efforts require special attention and priority.*

CRISIS RESPONSE

The transition from monitored volcanic activity to a volcanic crisis has as much to do with potential societal impact as with the nature of the eruptive phenomena. Within the United States, the USGS is expressly and uniquely empowered by the Stafford Act (Public Law 93-288) to issue timely warning of potential volcanic disasters to affected communities and civil authorities. Although not an explicitly mandated part of its mission, the VHP has also developed an international crisis response capability, the Volcano Disaster Assistance Program.

Often, the most valuable asset for a scientist responding to a crisis is relevant prior experience. Because its members are exposed to a wide variety of eruption styles and settings, VDAP offers the most effective way to prepare VHP staff for future domestic crises. The present system for selecting non-VDAP members of the VHP to join foreign deployments appears too haphazard. **The VHP should implement a more formal mechanism for participation in VDAP to see that as many people as possible are exposed to this type of training.**

Another missed opportunity for expanding the training potential of foreign volcanic crisis responses comes from the inability of VDAP members to archive their observations. *The success of VDAP should be measured not only in terms of mitigation of eruption impact, but also in terms of how well information and knowledge are disseminated in anticipation of future crises.* This change of strategy might ensure greater access to data that could be used to prepare future crisis teams.

A related programmatic issue is how staff members balance their responsibilities. Even if assistance were provided for archiving and distributing data from volcanic crises, individual scientists still have to incorporate their experiences into the published scientific literature. *The stated VHP goal of carefully documenting actual volcanic crises and responses is extremely important if the maximum information is to be obtained from any given eruption and is strongly endorsed by the committee.* This issue demands close monitoring, coordination, and allocation of staff time by the relevant scientists in charge to ensure that such information is forthcoming.

In addition to the valuable staff training opportunities provided by VDAP missions, foreign responses also allow new hardware and software to be evaluated under crisis conditions. The technical development of new instrumentation requires field tests for accurate calibration. A consequence of continuing tight VHP budgets has been the growing obsolescence of much of the equipment used in crisis response. One way in which the VHP can extend its equipment budget is to partner with manufacturers and other government agencies that design new instruments. **The committee encourages the VHP and VDAP to work more closely with NASA, DOE, DOD, and NOAA, as well as with NSF-funded consortia such as UNAVCO and IRIS, in the development of new in-**

strumentation and approaches suitable for detecting the conditions within erupting volcanoes.

The current level of VDAP funding allows a maximum of one deployment at a time, leading to occasional difficult decisions about priorities when multiple crises occur almost simultaneously. *The committee unconditionally supports the stated VHP desire to expand the size of the VDAP.*

PROGRAMMATIC AND INSTITUTIONAL ISSUES

Currently the VHP has a large number of capable scientists. However, *the almost total failure of the program to hire more than a token number of new personnel over the past 15 years has created a crisis of continuity in which much of the VHP's accumulated knowledge is in danger of being lost because of upcoming retirements.* Overlap of new staff with existing staff is essential for orderly transition of duties and transfer of knowledge, not only of volcanology and associated hydrology, but also of procedures for communicating with users of information. With the loss of personnel and no replacements, the domestic response capability is likely to collapse and programs such as VDAP could disappear. **The committee believes that if the VHP does not begin to hire new staff immediately, the program will not be able to maintain response readiness.** The committee suggests that the VHP begin planning for rejuvenation of its work force. This exercise should build upon the program's strategic plan and should take into account the new areas of expertise that will be needed in the future.

The importance of technicians to the VHP in many ways equals that of scientists. These individuals have highly eclectic backgrounds and in many cases have participated in several decades' worth of crisis response, especially as VDAP has expanded. *The lack of hiring in this area seriously threatens the well-being of the program.* Even if the number of VHP employees increases over the next few years, it will probably be insufficient to keep up with new techniques and with the increased flow of scientific knowledge that threatens to overwhelm the already overworked VHP staff. The resulting shortage means that the program will have to either reduce the scope of its mission or increase the pool of workers who can help them accomplish their goals. Because of this situation, *the committee concludes that the VHP can no longer accomplish all of its goals through in-house activities.* **The committee recommends that to accomplish its goals, the VHP increase its coordination and collaboration with researchers from other parts of the USGS, other federal agencies, academic institutions, and industry.** *The committee concludes that there is insufficient integration and communication between the VHP and other government entities involved in volcano hazards.* The VHP should take steps to ensure that USGS management realizes that the overall scientific goals of the program would be enhanced by such interactions. **The committee recommends that the VHP improve outside communication and better integrate its programs with those of other relevant organizations and government agencies.** One place where this coordination appears to be working well is in the separation between the assessment of volcanic hazards carried out by the VHP and the development of responses to those hazards conducted by local civil defense officials.

The VHP's Five-Year Science Plan for 1999 to 2003 outlines a wide array of program activities, ranging from volcano monitoring and crisis response to scientific outreach and information dissemination. If the VHP continues to be faced with flat budgets and limited staff growth, it must prioritize more clearly among these activities and see that they are consistent with stated program goals. *The committee urges the VHP to put in place a more formal mechanism for prioritizing its activities and seeing that they are consistent with stated program goals.*

Because most staff members of the VHP report to one of the scientists in charge of the four volcano observatories, these four individuals have special responsibilities for setting, assessing, enforcing, and coordinating prioritization across the program. In the observatory environment, volcano monitoring, hazard assessment, and communication with civil authorities may be most important, but during periods of volcano unrest and newly evolving activity, volcano crisis response assumes special priority.

A major issue that underlies any discussion of VHP priority setting and accountability is the lack of a clear and consistent management structure. Depending on his or her location and their inclination, an individual VHP scientist or technician might report to one of the four observatory scientists in charge, to the head of the Western Region in Menlo Park, to the local branch chief in Flagstaff, to the VHP coordinator in Reston, or to one of various administrators within the

Water Resources Division. The main drawback of the current complex structure is that it creates an institutional barrier to the emergence of strong leaders. This lack, in turn, makes individual staff members unsure about who sets their priorities. and makes the VHP as a whole less influential within the prioritization and budget-setting processes of the USGS and the Department of the Interior.

An important aspect of priority setting relates to the timeliness of scientific publication. Scientific publication is an important end product of VHP research, not only for the needs of civil authorities but also for other scientists (both USGS and non-USGS) who benefit from additions to the literature on volcanoes and volcano products. The problem is particularly acute when unpublished studies involve volcano hazard assessments that could have a direct bearing on the safety of people and property. *The committee urges that high priority be given to the timeliness of scientific publication.*

From the late 1960s until Mount St. Helens erupted in May 1980, the GD administrated the VHP and carried out all programmatic investigations. Soon after the Mount St. Helens event, the VHP funded a number of WRD projects, and the two divisions worked together as a single team. In the 1980s, disagreements between the two divisions prompted the USGS director to partition the VHP into two parts. This division in effect created two programs, each staffed and operated separately, based on different floors of the same building. It is questionable whether the scientific investigations and results throughout the program are integrated as effectively as they could be. The VHP is a USGS program and should be operated in ways that foster seamless relationships among staff within the GD and WRD. **The committee recommends that USGS management integrate the GD and WRD parts of the VHP.**

Standardization of data management protocols and formats across observatories and VDAP deployments is essential to improve access for the scientific community and others. The committee believes that the potential benefits of public access outweigh the possible drawbacks of data misuse. **The committee recommends that the VHP set standards for documentation, archiving, and access policies, including the length of the proprietary period.**

References

Eichelberger, J. 1997. Drilling volcanoes. Science 27:1084-1084.

Eichelberger, J. and Sattler, A. 1994. Conflict of values necessitates public lands research policy. EOS (Trans. Amer. Geophys. Union) 75(43):505-508.

Fink, J. H., and Griffiths, R. W. 1998. Morphology, eruption rates, and rheology of lava domes: Insights from laboratory models. Journal of Geophysical Research 103:527-545.

Hill, B. E., Connor, C. B., Jarzemba, M. S., La Femina P. C., Navarro, M., Strauch, W. 1998. 1995 eruptions of Cerro Negro volcano, Nicaragua, and risk assessment for future eruptions. Geological Society of America Bulletin 10:1231-1241.

Iverson, R. M. 1997. The physics of debris flows. Reviews of Geophysics 35:245-296.

Iverson, R. M., Schilling, S. P., and Vallance, J. W. 1998. Objective delineation of lahar-inundation hazard zones. Geological Society of America Bulletin 1108:972-984.

Malin, M. C. and Sheridan, M. F. 1982. Computer-assisted mapping of pyroclastic surges. Science 217:637-640.

Myers, B., Brantley, S. R., Stauffer, P., and Hendley, II J. W. 1997. What Are Volcano Hazards? Fact Sheet 002-97. Department of the Interior, U.S. Geological Survey. Washington, D.C.

Newhall C., Hendley, II J. W., Stauffer, P. H. 1997. Benefits of Volcano Monitoring Far Outweigh Costs—The Case of Mount Pinatubo. Fact Sheet 115-97. Department of the Interior, U.S. Geological Survey. Washington, D.C.

National Research Council (NRC). 1990. Letter report to Dr. Dallas Peck, Director, U.S. Geological Survey. Commission on Physical Science, Mathematics, and Resources, Washington, D.C.

Sheridan, M. F. 1979. Emplacement of pyroclastic flows: A review. Geological Society of America Special Paper 180:125-136.

Shoemaker E. M. Bredenhoeft, J. D., Christiansen, R. L., Gerlach, T. M., and Voight, B. 1986. Review of the Volcano Hazards Program of the U.S. Geological Survey. Reston, Va: U.S. Government Printing Office.

Simkin, T., Siebert, L., and Kimberly, P. 2000. Earth's volcanoes and eruptions: An overview. Encyclopedia of Volcanoes. San Diego, Ca. Academic Press.

Sparks, R. S. J., Bursik, M. I., Carey, S. N., Gilbert, J. S., Glaze, L. S. Sigurdsson, H., and Woods, A. W. 1997. Volcanic Plumes, New York: John Wiley & Sons.

Sutton, J., Elias, T., Hendley, II J. W., and Stauffer, P. H. 1997. Volcanic Air Pollution—A Hazard in Hawaii. Fact Sheet 169-97. Department of the Interior, U.S. Geological Survey. Washington, D.C.

Suzuki, T., Shimozuru, D., and Yokiyama, I., eds. 1983. A theoretical model for the dispersion of tephra. Arc Volcanism: Physics and Tectonics. Pp. 95-113. Japan: Terra Scientific Publishing Company.

U.S. Geological Suvey (USGS). 1992. Living with Volcanoes. Circular 1073. Reston, Va: U.S. Government Printing Office.

USGS. 1997. Five-Year Science Plan: 1998-2002. Reston, Va: U.S. Government Printing Office.

USGS. 1998. What Are Volcano Hazards? Fact Sheet 002-97. U.S. Geological Survey, Reston, Va: U.S. Government Printing Office.

USGS. 1999. Priorities for the Volcano Hazards Program 1999-2003. Reston, Va: U.S. Government Printing Office.

Wicks, Jr., C., Thatcher, W., and Dzurisin, D. 1998. Migration of fluids beneath Yellowstone caldera inferred from satellite radar interferometry. Science 282:458-672.

Appendixes

Appendix A
Biographical Sketches of Committee Members

Jonathan H. Fink (Chair) is vice provost for research and professor of geology at Arizona State University. His research focuses on the rheology, geochemistry, hazards, and emplacement of magma in environments ranging from the sea floor to the surfaces of other planets. He was Director of the Petrology and Geochemistry Program at NSF (1992-1993) and has served on several NSF review panels. He has been an editor of *Bulletin of Volcanology, Journal of Volcanology and Geothermal Research, and Journal of Geophysical Research*, and is a fellow of the Geological Society of America. In 1985, to expand the use of electronic communication among the volcanological community, Fink started Volcano Listserv, which now has more than 2,000 subscribers and is the principal source of volcanic information for scientists, journalists, policy makers, and other interested observers around the world.

Charles B. Connor is a principal scientist at the Center for Nuclear Waste Regulatory Analyses at Southwest Research Institute. His research interests include probabilistic analysis of volcanic hazards, geophysics of volcanoes, and mass and heat transfer processes on active volcanoes. His previous positions include associate professor of geology at Florida International University. Currently, he is a consultant on volcanic hazards to the International Atomic Energy Agency for development of agency guidelines on volcanic hazards and analysis of volcanic hazards at nuclear facilities in Armenia and Indonesia. He is member of the Science Committee for Colima Volcano, Mexico, and a member of the American Geophysical Union.

W. Gary Ernst is a professor of geological and environmental sciences at Stanford University, where he has been teaching for 10 years. During the previous 30 years he was professor of geology and geophysics at the UCLA. Ernst studies the deep-seated cores of Circumpacific and Alpine mountain belts as well as intracontinental suture zones in east-central China, the south Urals, and northern Kazakhstan. He investigates the subsolidus recrystallization of rocks during subduction-zone metamorphism and subsequent exhumation. A trustee of the Carnegie Institution of Washington, Ernst is also a member of the National Academy of Sciences, the American Academy of Arts and Sciences, and the American Philosophical Society. He served as president of the Mineralogical Society of America (1980-1981) and the Geological Society of America (1985-1986).

Richard S. Fiske is a geologist at the Smithsonian Institution's National Museum of Natural History. His research interests include submarine pyroclastic volcanism south of Japan and the history of explosive eruptions at Kilauea volcano, Hawaii. His previous positions include terms as director of the National Museum of Natural History and chief of the USGS Office of Geochemistry and Geophysics. He is fellow of the Geological Society of America and the American Advancement for the Association of Science and is a member of the American Geophysical Union.

Catherine J. Hickson is subdivision head of the Geological Survey of Canada's Vancouver office, Vancouver British Columbia, Canada. Additionally she manages a large multinational geoscience project in South America on behalf of the Canadian government. Her research interests include volcanism (especially subglacial volcanism), geological hazards, regional mapping, and emergency preparedness. She has a strong interest in scientific administration, international relations, and public education. She is a fellow of the Geological Association of Canada and the Geological Society of America in addition to membership in a number of other learned societies.

Harry Kim is administrator of the County of Hawaii Civil Defense Agency. A social scientist by training, his current duties are to administrate civil defense responsibilities as defined by the requirements of federal, state, and local governments. He has broad background and practical experience in emergency planning, emergency response, liability issues in emergency management, and risk communication. In the 1980s, he collaborated in the development of a volcanic emergency management handbook for the Office of the United Nations Disaster Relief. In recent years, he has played a major role in enhancing public awareness of the hazard of volcanic air pollution in Hawaii.

Stuart A. Rojstaczer is director of the Center for Hydrologic Science and associate professor of geology, environment and engineering at Duke

University. His research interest is to examine a wide range of hydrologic issues—some societally relevant, others of pure intellectual value—in which groundwater plays an essential role. He has held positions at the U.S. Geological Survey, Venice International University and is a visiting scientist at the Carnegie Institution of Washington.

Paul Segall is a professor of geophysics at Stanford University. His research interests include earthquake and volcano deformation, inversion of crustal deformation data, the mechanics of faulting, and the global positioning system. He is a fellow of the American Geophysical Union and Geological Society of America. He is presently using GPS to monitor deformation of Kilauea volcano in Hawaii and precise gravity changes to bound the density of magma intruding beneath Long Valley caldera, California. He is currently a member of the USGS Science Advisory Team for Long Valley caldera, the Southern California Integrated GPS Network Advisory Board, and the NSF Instruments and Facilities Program Panel.

John Stix is an associate professor of volcanology at McGill University. His research includes the investigation of shallow magmatic processes beneath active volcanoes by geological, geochemical, and geophysical means. He currently has projects at Long Valley caldera in California, Masaya volcano in Nicaragua, and Guagua Pichincha volcano in Ecuador. He is a member of the American Geophysical Union, the Geological Association of Canada, and the Geological Society of America.

Frederick J. Swanson is a research geologist with the Pacific Northwest Research Station of the U. S. Department of Agriculture Forest Service and Professor (Courtesy) in the Departments of Forest Science and Geoscience, Oregon State University. His research interests include interactions of forest and stream ecosystems with geophysical processes, such as those associated with volcanoes, floods, earthquakes, and wind storms. He also is involved in translation of findings from ecosystem research to management of forest lands and watersheds. For 12 years, he has been principal investigator for the Andrews Experimental Forest's Long-Term Ecological Research Program, sponsored by the National Science Foundation.

Tamara L. Dickinson (staff) is a senior program officer for the Board on Earth Sciences and Resources of the National Research Council. She has served as program director for the Petrology and Geochemistry

Program in the Division of Earth Sciences at the National Science Foundation. She has also served as discipline scientist for the Planetary Materials and Geochemistry Program at NASA Headquarters. As a post-doctoral fellow at the NASA Johnson Space Center, she conducted experiments on the origin and evolution of lunar rocks and highly reduced igneous meteorites. She holds a Ph.D. and a M.S. in geology from the University of New Mexico and a B.A. in geology from the University of Northern Iowa.

Rebecca E. Shapack (staff) is a research assistant for the Board on Earth Sciences and Resources of the National Research Council. She holds a B.S. in mathematical sciences engineering with a concentration in biology from the Johns Hopkins University, and is currently working on her M.S. in public health at the George Washington University.

Appendix B
Oral Presentations and Written Statements

ORAL PRESENTATIONS

The following individuals made presentations to the committee on the Review of the USGS Volcano Hazards Program:

USGS *Charles Groat, Director.*

USGS GD *Harley M. Benz, Earthquake Program, Denver; P. Patrick Leahy, Chief Geologist.*

USGS WRD *Robert Hirsch, Chief Hydrologist; Michael Sorey, National Research Program.*

USGS VHP *Steven R. Brantley, Outreach Coordinator, Hawaii Volcano Observatory; Phillip B. Dawson, Project Scientist, Menlo Park; Carolyn L. Driedger, Coordinator of Educational Outreach Programs, Water Resources Division, Cascades Volcano Observatory; Daniel Dzurisin, Geologist, Cascades Volcano Observatory; John W. Ewert, Volcano Disaster Assistance Program, Cascades Volcano Observatory; Marianne Guffanti, Coordinator, Volcano Hazards Program, Reston; Edward W. Hildreth, Geologist, Volcano Hazards Program, Menlo Park; David P. Hill, Scientist in Charge, Long Valley Observatory; Richard M. Iverson, Water Resources Division, Cascades Volcano Observatory; Terry E. Keith, Scientist in Charge, Alaska Volcano Observatory; Richard G. LaHusen, Hydrologist, Water Resources Division, Cascades Volcano Observatory; Michael Lisowski, Deformation Specialist, Hawaii Volcano Observatory; C. Daniel Miller, Chief, Volcano Disaster Assistance Program, Cascades Volcano Observatory; Thomas L. Murray, Geophysicist, Cascades Volcano Observatory; Thomas C. Pierson, Associate Scientist in Charge, Water Resource Division, Cascades Volcano Observatory; John A. Power, Geophysicist, Volcano Disaster Assistance Program Alaska Volcano Observatory; Steve P. Schilling, Hydrologist, Cascades Volcano Observatory; William E. Scott, Scientist* *in Charge Cascades Volcano Observatory; Donald A. Swanson, Scientist in Charge, Hawaiian Volcanoes Observatory; Robert Tilling, Chief Scientist, Menlo Park; Christopher F. Waythomas, Project Director, Water Resources Division, Alaska Volcano Observatory.*

Other Federal Government Perspectives

Donald "Doc" Carver, Assistant Federal Coordinator for Transportation, Federal Aviation Administration; Arlin Krueger, NASA Goddard Space Center, James Luhr, Director of the Global Volcanism Program, Smithsonian Natural History Museum; James Martin, Superintendent, Hawaii Volcanoes National Park; Raymond Meyer, Deputy Assistant Director for Technical Assistance, Office of Foreign Disaster Assistance; David Russell, Legislative Assistant, Senator Stevens Office (AK-R);Grace Swanson, Staff Meteorologist, National Oceanic and Atmospheric Administration, Lou Walter, Program Manager, Solid Earth Sciences and Natural Hazards, NASA Headquarters.

Other Public and Private Perspectives

Steve Bailey, Department of Emergency Management, Peirce County, Washington; Katharine Cashman, Professor, Department of Geological Sciences, University of Oregon; John Eichelberger, Professor of Volcanology, University of Alaska, and Coordinating Scientist, Alaska Volcano Observatory; Tracy Fuller, Town Manager, Mammoth Lakes, California; Stephen Malone, Research Professor, Graduate Program in Geophysics, University of Washington; Peter Mouginis-Mark, Associate Director, Hawaii Institute of Geophysics and Plantology, University of Hawaii, Acting Federal Program Scientist, Pacific Disaster Center Kihei, Maui;Captain Edward Miller, Project Leader, Volcanic Ash and Aviation Safety Project, Aviation Weather Committee, Airline Pilots Association; J. Bernard Minster, Professor, University of California, San Diego, Scripps Institution of Oceanography; Christopher G. Newhall, Associate Professor, University of Washington; David Pieri, Earth Space and Science Division, Jet Propulsion Laboratory; Dave Rider, Logistics Coordinator, Washington Military Department, Washington; Vince

Realmuto, Supervisor, Visualization and Earth Science Applications Group, Jet Propulsion Laboratory; Boe Turner, Office of Emergency Management, Mono County Sheriffs Office, Mammoth Lakes, California; David Unnewehr, Senior Research Manager, American Insurance Association.

WRITTEN STATEMENTS

The following individuals provided written statements to the panel either at the request of the panel or as unsolicited input:

Minard L. Hall, Instituto Geofisico; Lindsay McClelland, National Park Service; Peter Mouginis-Mark, University of Hawaii; William Rose, Michigan Technological University; Stephen Self, University of Hawaii; Stephen Sparks, University of Bristol; Barry Voight, Pennsylvania State University; Simon Young, Montserrat Volcano Observatory.

Appendix C

USGS Volcano Hazard Assessments[*]

Multi State

Mullineaux, D. R. 1976. Preliminary overview map of volcanic hazards in the 48 conterminous United States. U.S. Geological Survey Miscellaneous Field Studies Map MF-786, 1 plate, scale 1:7,500,000.

Shipley, S., and Sarna-Wojcicki, A. M. 1983. Distribution, thickness, and mass of late Pleistocene and Holocene tephra from major volcanoes in the northwestern United States: A preliminary assessment of hazards from volcanic ejecta to nuclear reactors in the Pacific northwest. U.S. Geological Survey Miscellaneous Field Studies Map MF-1435, 27 Pp., 1 plate, scale 1:2,500,000.

Hoblitt, R. P., Miller, C. D., and Scott, W. E. 1987. Volcanic hazards with regard to siting nuclear-power plants in the Pacific Northwest. U.S. Geological Survey Open-File Report 87-297, 196 Pp., 5 plates, scale 1:2,000,000.

Alaska

Akutan

Waythomas, C. F., Power, J. A., Richter, D. H., and McGimsey, R. G. 1998. Preliminary volcano-hazard assessment for Akutan Volcano east-central Aleutian Islands, Alaska. U.S. Geological Survey Open-File Report 98-360, 36 Pp., 1 plate.

---

[*] As of September 1999

Augustine
   Waythomas, C. F., and Waitt, R. B. 1998. Preliminary volcano-hazard
      assessment for Augustine Volcano, Alaska. U.S. Geological
      Survey Open-File Report 98-106, 39 Pp., 1 plate.

Iliamna
   Waythomas, C. F., and Miller, T. P. 1999. Preliminary volcano hazard
      assessment for Iliamna Volcano, Alaska. U.S. Geological Survey
      Open-File Report 99-373, 31 Pp., 1 plate.

Pavlof
   Waythomas, C. F., Miller, T. P., McGimsey, R. G., and Neal, C. A.
      1997. Preliminary volcano-hazard assessment for Pavlof Volcano,
      Alaska. U.S. Geological Survey Open-File Report 97-135, 1
      plate.

Redoubt
   Till, A. B., Yount, M. E., Riehle, J. R. 1993. Redoubt Volcano,
      southern Alaska: A hazard assessment based on eruptive activity
      through 1968. U.S. Geological Survey Bulletin 1996, 19 Pp., 1
      plate, scale 1:125,000.
   Waythomas, C. F., Dorava, J. M., Miller, T. P., Neal, C. A., and
      McGimsey, R. G. 1998. Preliminary volcano-hazard assessment
      for Redoubt Volcano, Alaska. U.S. Geological Survey Open-File
      Report 97-857, 40 Pp., 1 plate.

Hawaii

Island of Hawaii
   Mullineaux, D. R., and Peterson, D. W. 1974. Volcanic hazards on the
      island of Hawaii. U.S. Geological Survey Open-File Report 74-
      239, 61 Pp., 1 plate, scale 1:125,000.
   U.S. Geological Survey (from material provided by Peterson, D. W.,
      and Mullineaux, D. R.). 1976. Natural hazards on the island of
      Hawaii. U.S. Geological Survey INF-75-18, 16 Pp.
   Mullineaux, D. R., Peterson, D. W., and Crandell, D. R. 1987.
      Volcanic hazards in the Hawaiian Islands, Pp. 599-621 in Decker,

R. W., Wright, T. L., and Stauffer, P. H., eds., Volcanism in Hawaii. U.S. Geological Survey Professional Paper 1350, v. 1.

Heliker, C. 1990. Volcanic and seismic hazards on the Island of Hawaii. U.S. Geological Survey General Interest Publication, 48 Pp.

Wright, T. L., Chun, J. Y. F., Esposo, J., Heliker, C., Hodge, J., Lockwood, J. P., and Vogt, S. M. 1992. Map showing lava-flow hazard zones, Island of Hawaii. U.S. Geological Survey Miscellaneous Field Studies Map MF-2193, 1 plate, scale 1:250,000.

Kauahikaua, J., Moore, R. B., and Delaney, P. 1994. Volcanic activity and ground deformation hazard analysis for the Hawaii Geothermal Project Environmental Impact Statement. U.S. Geological Survey Open-File Report 94-553, 44 Pp.

Kauahikaua, J., Margriter, S., and Moore, R. B. 1995. GIS-aided volcanic activity hazard analysis for the Hawaii Geothermal Project Environmental Impact Statement, Pp. 235-257 in Carrara, A., and Guzzetti, F., eds., Geographical Information Systems in Assessing Natural Hazards, Amsterdam: Kluwer.

Mauna Loa

Kauahikaua, J., Margriter, S., Lockwood, J., and Trusdell, F. 1995. Applications of GIS to the estimation of lava flow hazards on Mauna Loa Volcano, Hawaii Pp. 315-325 in Rhodes, J. M., and Lockwood, J. P., eds., Mauna Loa Revealed; Structure, Composition, History, and Hazards. American Geophysical Union, Geophysical Monograph 92.

Trusdell, F. A. 1995. Lava flow hazards and risk assessment on Mauna Loa Volcano, Hawaii, Pp. 327-336 in Rhodes, J. M., and Lockwood, J. P., eds., Mauna Loa Revealed; Structure, Composition, History, and Hazards. American Geophysical Union, Geophysical Monograph 92.

Kauahikaua, J., Trusdell, F., and Heliker, C. 1998. The probability of lava inundation at the proposed and existing Kulani Prison sites. U.S. Geological Survey Open-File Report 98-794, 21 Pp.

Oahu
Crandell, D. R. 1975. Assessment of volcanic risk on the Island of Oahu, Hawaii. U.S. Geological Survey Open-File Report 75-287, 18 Pp.

Maui
Crandell, D. R. 1983. Potential hazards from future volcanic eruptions on the Island of Maui, Hawaii. U.S. Geological Survey Miscellaneous Investigations Map I-1442, scale 1:100,000.

Haleakala
Mullineaux, D. R., Peterson, D. W., and Crandell, D. R. 1987. Volcanic hazards in the Hawaiian Islands, Pp. 599-621 in Decker, R. W., Wright, T. L., and Stauffer, P. H., eds., Volcanism in Hawaii. U.S. Geological Survey Professional Paper 1350, v. 1.

Washington

General
Crandell, D. R. 1976. Preliminary assessment of potential hazards from future volcanic eruptions in Washington. U.S. Geological Survey Miscellaneous Field Studies Map MF-774, 1 plate, scale 1:1,000,000.

Glacier Peak
Begét, J. E. 1982. Postglacial volcanic deposits at Glacier Peak, Washington, and potential hazards from future eruptions. U.S. Geological Survey Open-File Report 82-830, 77 Pp.
Waitt, R. B., Mastin, L. G., and Begét, J. E. 1995. Volcanic-hazard zonation for Glacier Peak Volcano, Washington. U.S. Geological Survey Open-File Report 95-499, 9 Pp., 2 plates, scale 1:100,000.

Mount Adams
Scott, W. E., Iverson, R. M., Vallance, J. W., and Hildreth, W. 1995. Volcano hazards in the Mount Adams region, Washington. U.S. Geological Survey Open-File Report 95-492, 11 Pp., 2 plates, scale 1:500,000, 1:100,000.

Mount Baker
Hyde, J. H., and Crandell, D. R. 1978. Postglacial volcanic deposits at Mount Baker, Washington, and potential hazards from future eruptions. U.S. Geological Survey Professional Paper 1022-C, 17 Pp., 1 plate, scale 1:250,000.
Gardner, C. A., Scott, K. M., Miller, C. D., Myers, B., Hildreth, W., and Pringle, P. T. 1995. Potential volcanic hazards from future activity of Mount Baker, Washington. U.S. Geological Survey Open-File Report 95-498, 16 Pp., 1 plate, scale 1:100,000.

Mount Rainier
Crandell, D. R., and Mullineaux, D. R. 1967. Volcanic hazards at Mount Rainier, Washington. U.S. Geological Survey Bulletin 1238, 26 Pp.
Crandell, D. R. 1973. Potential hazards from future eruptions of Mount Rainier, Washington. U.S. Geological Survey Miscellaneous Geological Investigations Map I-836, 1 plate, scale 1:250,000.
Hoblitt, R. P., Walder, J. S., Driedger, C. L., Scott, K. M., Pringle, P. T., and Vallance, J. W. 1995. Volcano hazards from Mount Rainier, Washington. U.S. Geological Survey Open-File Report 95-273, 10 Pp., 1 plate, scale 1:100,000.
Scott, K. M., and Vallance, J. W. 1995. Debris flow, debris avalanche, and flood hazards at and downstream from Mount Rainier, Washington. U.S. Geological Survey Hydrologic Investigations Atlas HA-729, 9 Pp., 1 plate, scale 1:100,000, 1:400,000.
Scott, K. M., Vallance, J. W., and Pringle, P. T. 1995. Sedimentology, behavior, and hazards of debris flows at Mount Rainier, Washington. U.S. Geological Survey Professional Paper 1547, 56 Pp., 1 plate.
U.S. Geological Survey. 1996. Perilous Beauty, The Hidden Dangers of Mount Rainier. VHS video, 29 min.
Hoblitt, R. P., Walder, J. S., Driedger, C. L., Scott, K. M., Pringle, P. T., and Vallance, J. W. 1998. Volcano hazards from Mount Rainier, Washington, Revised 1998. U.S. Geological Survey Open-File Report 98-428, 11 Pp., 2 plates, scale 1:100,000, 1:400,000.

Mount St. Helens

Crandell, D. R. and Mullineaux, D. R. 1978. Potential hazards from future eruptions of Mount St. Helens Volcano, Washington. U.S. Geological Survey Bulletin 1383-C, 26 Pp., 2 plates, scale 1:250,000.

Miller, C. D., Mullineaux, D. R., and Crandell, D. R. 1981. Hazards assessments at Mount St. Helens, Pp. 789-802 in Lipman, P. W., and Mullineaux, D. R., eds., The 1980 Eruptions of Mount St. Helens, Washington. U.S. Geological Survey Professional Paper 1250.

Newhall, C. G. 1982. A method for estimating intermediate- and long-term risks from volcanic activity, with an example from Mount St. Helens, Washington. U.S. Geological Survey Open-File Report 82-396, 59 Pp.

Newhall, C. G. 1984. Semiquantitative assessment of changing volcanic risk at Mount St. Helens, Washington. U.S. Geological Survey Open-File Report 84-272, 30 Pp.

Pallister, J. S., Hoblitt, R. P., Crandell, D. R., and Mullineaux, D. R. 1992. Mount St. Helens a decade after the 1980 eruptions: Magmatic models, chemical cycles, and a revised hazards assessment. Bulletin of Volcanology, 54:126-146.

Wolfe, E. W., and Pierson, T. C. 1995. Volcanic-hazard zonation for Mount St. Helens, Washington, 1995. U.S. Geological Survey Open-File Report 95-497, 12 Pp., 1 plate, scale 1:100,000.

Oregon

Crater Lake

Bacon, C. R., Mastin, L. G., Scott, K. M., and Nathenson, M. 1997. Volcano and earthquake hazards in the Crater Lake region, Oregon. U.S. Geological Survey Open-File Report 97-487, 32 Pp., 1 plate, scale 1:100,000.

Mount Hood

Crandell, D. R. 1980. Recent eruptive history of Mount Hood, Oregon, and potential hazards from future eruptions. U.S. Geological Survey Bulletin 1492, 81 Pp., 1 plate, scale 1:31,680.

Scott, W. E., Pierson, T. C., Schilling, S. P., Costa, J. E., Gardner, C. A., Vallance, J. W., and Major, J. J. 1997. Volcano hazards in the Mount Hood region, Oregon. U.S. Geological Survey Open-File Report 97-89, 14 Pp., 1 plate, scale 1:100,000.

Wessells, Steve. 1998. At risk: Volcano Hazards from Mount Hood, Oregon. U.S. Geological Survey Open-File Report 98-492, VHS video, 14 min.

Mount Jefferson

Walder, J. S., Gardner, C. A., Conrey, R. M., Fisher, B. J., and Schilling, S. P. 1999. Volcano hazards in the Mount Jefferson region, Oregon. U.S. Geological Survey Open-File Report 99-24, 14 Pp., 2 plates, scale 1:100,000.

Newberry Volcano

Sherrod, D. R., Mastin, L. G., Scott, W. E., and Schilling, S. P. 1997. Volcano hazards at Newberry Volcano, Oregon. U.S. Geological Survey Open-File Report 97-513, 14 Pp., 1 plate, scale 1:100,000.

Three Sisters Volcano

Laenen, A., Scott, K. M., Costa, J. E., and Orzol, L. L. 1987. Hydrologic hazards along Squaw Creek from a hypothetical failure of the glacial moraine impounding Carver Lake near Sisters, Oregon. U.S. Geological Survey Open-File Report 87-41, 48 Pp.

Scott, W. E., Iverson, R. M., Schilling, S. P., and Fisher, B. J. 1999. Volcano hazards in the Three Sisters Region, Oregon. U.S. Geological Survey Open-File Report 99-437, 18 Pp., 1 plate, scale 1:150,000.

California

General

Christiansen, R. L. 1982. Volcanic hazard potential in the California Cascades, Pp. 41-59 in Martin, R. C., and Davis, J. F. Status of Volcanic Prediction and Emergency Response Capabilities in Volcanic Hazard Zones of California. California Division of Mines and Geology Special Publication 63.

Miller, C. D. 1989. Potential hazards from future volcanic eruptions in California. U.S. Geological Survey Bulletin, 1847, 17 Pp., 2 plates, scale 1:500,000.

Lassen Volcanic Center

Crandell, D. R., and Mullineaux, D. R. 1970. Potential geologic hazards in Lassen Volcanic National Park, California. U.S. Geological Survey Administrative Report, 54 Pp.

Mount Shasta

Miller, C. D. 1980. Potential hazards from future eruptions in the vicinity of Mount Shasta volcano, northern California. U.S. Geological Survey Bulletin 1503, 43 Pp., 3 plates, scale 1:62,500.

Crandell, D. R., and Nichols, D. R. 1987. Volcanic hazards at Mount Shasta, California. U.S. Geological Survey General Interest Publication, 21 Pp.

Crandell, D. R. 1989. Gigantic debris avalanche of Pleistocene age from ancestral Mount Shasta Volcano, California, and debris-avalanche hazard zonation. U.S. Geological Survey Bulletin 1861, 32 Pp.

Ostercamp, W. R., Hupp, C. R., and Blodgett, J. C. 1986. Magnitude and frequency of debris flows, and areas of hazards on Mount Shasta, northern California. U.S. Geological Survey Professional Paper 1396-C, 21 Pp., 1 plate, scale 1:62,500.

Long Valley Caldera

Miller, C. D., Mullineaux, D. R., Crandell, D. R., and Bailey, R. A. 1982. Potential hazards from future eruptions in the Long Valley–Mono Lake area, east-central California and southwest Nevada—a preliminary assessment. U.S. Geological Survey Circular 877, 10 Pp.

Foreign

Miller, C. D., Mullineaux, D. R., and Hall, M. L. 1978. Reconnaissance map of potential volcanic hazards from Cotopaxi Volcano, Ecuador. U.S. Geological Survey Miscellaneous Investigations Map I-1072, 1 plate, scale 1:100,000.

Miller, C. D., Sukhyar, R., Santoso, and Hamidi, S. 1983 Eruptive history of the Dieng Mountains region, central Java, and potential hazards from future eruptions. U.S. Geological Survey Open-File Report 83-68, 20 Pp.

Macías, V. J. L., Carrasco-Nunez, G., Delgado G. H., Martin del Pozzo, A. L., Siebe G., C., Hoblitt, R. P. Sheridan, M. F., and Tilling, R. I. 1995. Mapa de peligros del Volcan Popocatépetl. Instituto de Geofisica, Universidad Nacional Autonoma B Mexico, Mexico, D.F., 1 plate, scale 1:250,000.

Punongbayan, R. S., Newhall, C. G., Bautista, M. L. P., Garcia, D., Harlow, D. H., Hoblitt, R. P., Sabit, J. P., and Solidum, R. U. 1996. Eruption hazards assessments and warnings, in Newhall, C. G., and Punongbayan, R. S., eds., Fire and Mud: Eruptions and Lahars of Mount Pinatubo, Philippines. Philippine Institute of Volcanology and Seismology, Quezon City, and University of Washington Press, Seattle, Pp. 67-85

Acronyms

AFM	acoustic flow monitors
AVO	Alaska Volcano Observatory
CINCPAC	Commander-in-Chief, U.S. Pacific Command
CINDI	Center for Integration of Natural Disaster Information
CVO	Cascades Volcano Observatory
DOD	Department of Defense
DOE	Department of Energy
DOI	Department of the Interior
EOS	Earth Observing System
FAA	Federal Aviation Administration
FEMA	Federal Emergency Management Agency
FTE	full-time equivalent
FTIR	Fourier transform infrared spectroscopy
GASPEC	gas correlation spectroscopy
GD	Geologic Division
GDIN	Global Disaster Information Network
GIS	geographic information system
GOES	Geostationary Operational Environmental Satellites
GPS	global positioning system
HSS	Hazard Support System
HVO	Hawaiian Volcano Observatory
IAVCEI	International Association of Volcanology and Chemistry of the Earth's Interior
InSAR	interferometric synthetic aperture radar
IPA	Intergovernmental Personnel Act
IRIS	Incorporated Research Institutions for Seismology
LIDAR	light intensity detection and ranging
LTER	Long Term Ecological Research
LVO	Long Valley Observatory

NASA	National Aeronautics and Space Administration
NOAA	National Oceanic and Atmospheric Administration
NPS	National Park Service
NRC	National Research Council
NSF	National Science Foundation
NWS	National Weather Service
OFDA	U.S. Office of Foreign Disaster Assistance
PDC	Pacific Disaster Center
PHIVOLCS	Philippine Institute of Volcanology and Seismology
UNAVCO	University NAVSTAR Consortium
USAID	U.S. Agency for International Development
USGS	United States Geological Survey
VDAP	Volcano Disaster Assistance Program
VEI	Volcanic Explosivity Index
VHP	Volcano Hazards Program
WRD	Water Resources Division